IEEE Standard Digital Interface for Programmable Instrumentation

IEEE Recommended Practice for Code and Format Conventions for Use with ANSI/IEEE Std 488-1978

Introduction
by
Donald C. Loughry

Library of Congress Catalog Number 83-83197
ISBN 471-80786-9

The Institute of Electrical and Electronics Engineers, Inc
345 East 47th Street, New York, NY 10017

SH08748

IEEE Standard Digital Interface for Programmable Instrumentation

IEEE Recommended Practice for Code and Format Conventions for Use with ANSI/IEEE Std 488-1978

Introduction
by
Donald C. Loughry

An interface system specification optimized for the interconnection of programmable instrumentation must contain many vital communication capabilities if it is to interconnect independently manufactured products; at reasonable cost, with a high degree of compatibility, and flexibility to apply to a wide range of products. Two IEEE Standards have been developed to facilitate the design, manufacture, and use of instrumentation destined to be assembled in small systems;

ANSI/IEEE Std 488-1978, IEEE Standard Digital Interface for Programmable Instrumentation,

ANSI/IEEE Std 728-1982, IEEE Recommended Practice for Code and Format Conventions.

Each standard has been written as a reference document and as such, text of a tutorial nature has been omitted. The inclusion of text aimed at providing background material with rationale for why certain functions are included or excluded would have made the specifications too lengthy.

This introduction is intended to provide a very brief overview of the interface concepts specified throughout IEEE Std 488 and is considered helpful to readers unfamiliar with the standard. The reader may want to consult two other articles thought to present useful background material — see [1] and [2].[1]

[1] These references in brackets are listed on page 8.

Historical Perspective

IEEE Std 488-1975, *IEEE Standard Digital Interface for Programmable Instrumentation*, was elaborated to facilitate the design, assembly, and use of instrumentation systems. What conditions promoted and what organizations participated in the development of this standard?

Detailed discussion regarding the feasibility, scope, and objectives for a standard interface commenced in the United States and Europe at approximately the same time, mid-1972. Such discussions took place in response to the rising need to minimize, if not eliminate, the resources expended repeatedly on unique solutions to the interconnection of system components. Digital voltmeter, counter, and signal generator design engineers, for example, tended to select interface methods that satisfied specific and short term needs. The choice of codes, logic sense and levels, and signal and timing protocols were tailored to fit one primary application. Development and manufacturing costs were relatively high, each solution being somewhat unique. Once an instrument had some means to handle I/0 data, the inevitable happened. Customers, system manufacturers, and users of all types sought to assemble new system configurations beyond the scope envisioned for the initial interface. The end result, additional resources and special engineering were applied to make incompatible instruments communicate at a rudimentary level. These situations existed for those individuals configuring systems from different manufacturers and systems comprised of a single manufacturer's products. This condition was further heightened by the need for and trends toward automated systems with the obvious benefits of: greater throughput and accuracy, more thorough testing, results in engineering units, and lack of operator fatigue.

Clearly the need existed for a common solution, an interface standard applicable to a wide variety of products, an interface standard that would benefit manufacturers and users alike. Under the guidance of the International Electrotechnical Commission (IEC) and instigation of the German National Committee, a development project was authorized and Working Group 3 (IEC/TC66/WG3) formed in 1972. Concurrently a USNC advisory committee and IEEE subcommittee were formed to consider the needs of the US manufactures and users. At the first few IEEE meetings the overall interface objectives were formulated and several proposals evaluated. The interface concepts then under development at the Hewlett-Packard Company were selected as the model for further discussion and development. The first international WG3 meeting was held in Munich, Germany, October 1972. Again, after agreement on overall objectives, the US proposal was selected as the basis for elaboration of an international IEC standard.

Over the next two years the basic concepts were elaborated at the national and international levels and included such refinements as address extention, parallel poll capability, control shift, and remote-local enhancements. In December 1974, the IEEE Standards Board approved what is now known as IEEE Std 488. During this process, at the national and international levels, a significant contribution was made toward the clarity of the interface definition by the use of state diagram concepts and notations [2].

The IEEE Std 488-1975 was first published in April of 1975. Its international counterpart, IEC Publication 625-1, *An Interface System for Programmable Measuring Instruments, Byte Serial, Bit Parallel* was approved and published in 1979. Two further revisions to the basic IEEE Std 488 have taken place since the initial publication. First, the standard was revised in 1978 to incorporate some editorial clarifications and to remove potentially ambiguous statements. Specific details concerning this revision are included in the Foreword of the standard. In 1980, Supplement A was approved and has been incorporated in all subsequent printings of the standard. The purpose of the supplement was to further clarify the take control synchronous (tcs) conditions. The supplement adds a single state (CSHS, Controller Standby Hold State) to the controller function as defined in 2.12. The addition of this state and the associated transitions guarantees that idle devices will not falsely interpret messages during a synchronous take-control sequence.

Fundamental Interface Concepts

Interface standards are no different from other standards in that manageable objectives, limited scope, basic structures, and a realistic perspective of what is appropriate to standardize on, all aid in the elaboration of useful concepts. The scope of IEEE Std 488 focuses on three key parameters restricting its use to applications in which the system components are in close proximity to one another. In terms relevant to a communication link, the scope of IEEE Std 488 addresses the interconnection of up to 15 devices on one local network, message flow over a maximum transmission path length of 20 m, and data rates at a maximum of 1 Mb/s. If these values are considered as upper limits, then a further objective is to provide the user (design engineer and end user) with as much freedom as possible (for example, any number of devices, asynchronous data transfer, flexible and simple physical connection schema) within the boundary conditions.

Another useful perspective partitions an interface system into four main elements: mechanical, electrical, functional, and operational as shown in Fig 1. The first element, mechanical, pertains to the physical means of

Fig 1
Major Interface System Elements

MECHANICAL	ELECTRICAL	FUNCTIONAL	OPERATIONAL
SCOPE OF INTERFACE SYSTEM DEFINITION DEVICE (SYSTEM) INDEPENDENT SPECIFICATIONS			DEVICE (SYSTEM) DEPENDENT SPECIFICATIONS

interconnect, cables, and connectors. Voltage and current limits, logic conventions, and overall specifications for the interchange medium comprise the electrical aspects of an interface.

The functional interface elements relate to the precise definition of each signal line, its purpose and method of use, the protocol and related timing constraints required to exchange messages among interconnected devices. More costly to implement than either of the other two elements, the functional aspects of an interface have a significiant impact on compatibility between independently designed products. IEEE Std 488 devotes over two-thirds of the technical description to functional considerations. Each of these interface elements can be viewed and defined in a device and system in an independent manner. *The feasibility of an interface standard is increased significantly when it is devoid of characteristics and features aimed at one or a few product types. IEEE Std 488 defines the mechcanical, electrical, and functional specifications in a device and system independent manner.* The fourth element of an interface, operational characteristics, relates more to the device dependent parameters (for example, applications software program codes, diagnostic routines) and is not addressed in the IEEE Std 488 standard.

IEEE Std 488 Overview

The IEEE Std 488-1978 interface system utilizes a party-line bus structure to which a maximum of 15 devices may be connected. Sixteen signal lines (see Fig 2) provide the functional capability necessary to maintain an orderly flow of device and interface related messages.

To describe further the interface performance, it is convenient to use terms appropriate for communication between human beings, talkers, and listeners. In the bus context, a "talker" implies that the device is enabled to send data over the bus. Similarly "listener(s)" are those devices enabled to receive data over the bus. In addition, a "controller" is required to designate which device is to talk and which device(s) are to listen. A controller also is capable of sending special types of data (for example, Universal commands) to all devices connected to the bus and receiving status data from other devices.

Eight bidirectional data lines, D101-D108, carry a variety of coded messages in byte-serial bit-parallel form across the interface. Typical coded messages include addresses, program data to set up or configure a programmable instrument, measurement results, universal commends of interest to all devices,

Fig 2
Party-line Bus Structure Services Fifteen Devices

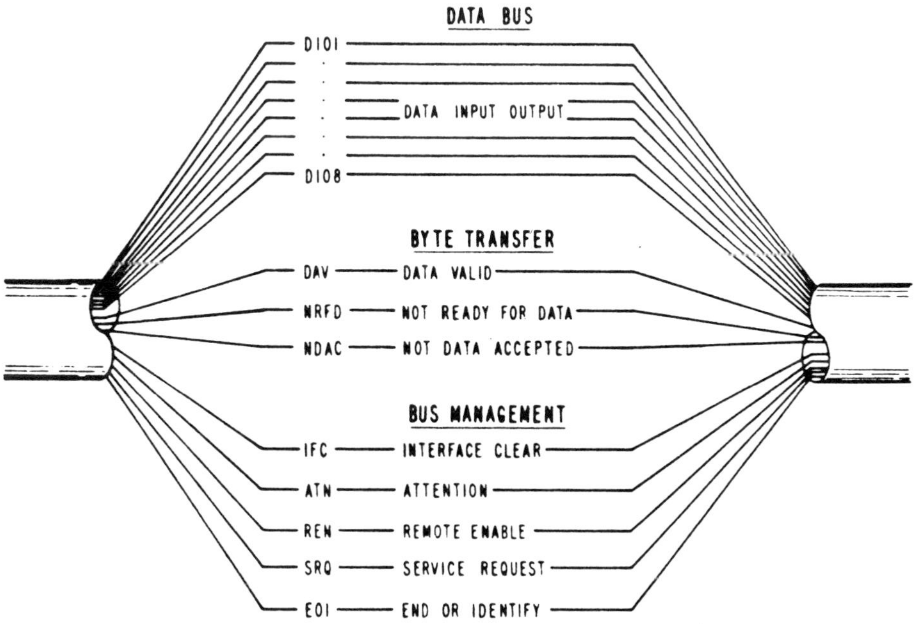

and solicited status data. Identification of the type of data present on the DIO signal lines is achieved by the ATN (attention) message. An ATN message, sent true by a controller, indicates that the DIO signal lines contain addresses or universal commands of interest to all devices. When the ATN message is sent false, only those devices addressed to talk or listen communicate by way of DI01-DI08 and exchange device-dependent data.

Data byte transfer on the DIO signal lines is accomplished using an interlocked handshake sequence on three signal lines; data valid (DAV), not ready for data (NRFD) and not data accepted (NDAC). The talker utilizes DAV to signal the presence of a valid byte of data on DI01-DI08. Each listener indicates its acceptance of the data byte by asserting the RFD message false and the DAC message true. If only one device on the bus is an active listener, the existence of the DAC message alerts the talker that the data byte has been received and is no longer needed. The talker then assets the DAV message false and the listener, when it is able to receive a new data byte, asserts RFD true and DAC false. Thus the handshake sequence is complete. When two or more devices are "listening" (for example, address mode or DVM output to printer and controller) each device participates in the same handshake cycle. The major difference, however, is that each step of the handshake cycle is not complete until all devices have acted upon each step in the process. Due to the bus structure and the use of open collector drivers on the NRFD and NDAC signal lines, a logical AND (wired OR) function is achieved. The RFD message is not seen by the talker until *all* listeners are ready for data, the DAC message is not seen by the talker until the *last* or slowest of the devices has accepted the data. This process enables devices with different data acceptance rates to coexist on the same bus. *Data is thus transferred asynchronously at the rate of the slowest listener participating in the conversation. High-speed data transfer between two devices is not adversely affected by low speed devices physically connected to the bus whenever these slower devices are not addressed to talk or listen.*

The interface clear (IFC) message is used by a controller to place the interface system in a known quiescent state. Devices are able to indicate to the controller summary status data by means of the service request (SQR) message. Assertion of the SRQ message does not mandate immediate interruption of the current sequence of events, the controller is free to respond when approporiate. Instrumentation systems may require a mixture of local and remote control. This capability is achieved, in part, by the remote enable (REN) signal line used to select between local front panel controls and remote control by way of the interface bus. The end or identify (EOI) signal line serves two purposes. First, it is used to the active talker to send the END message indicating the last byte of a multibyte data transfer. The EOI signal line may be used by the controller to execute a parallel poll (single status bit from each of eight devices concurrently) when the identify (IDY) message is asserted concurrently with the ATN message. Thus the 16 signal line bus structure enables an orderly flow of messages among interconnected devices.

Table 1
Typical Remote Messages Defined in IEEE Std 488

REMOTE MESSAGE		CLASS	CODING									
			DIO								ATN	SRQ
			8	7	6	5	4	3	2	1		
DAB	data byte	DD	D	D	D	D	D	D	D	D	0	X
DCL	device clear	UC	X	0	0	1	0	1	0	0	1	X
MLA	my listen address	AD	X	0	1	L	L	L	L	L	1	X
MTA	my talk address	AD	X	1	0	T	T	T	T	T	1	X
RQS	request service	ST	X	1	X	X	X	X	X	X	0	X
SPE	serial poll enable	UC	X	0	0	1	1	0	0	0	1	X
SRO	service request	ST	X	X	X	X	X	X	X	X	X	1
STB	status byte	ST	S	X	S	S	S	S	S	S	X	X
UNL	unlisten	AD	X	1	0	1	1	1	1	1	1	X

Coding: X = don't care, 1 = true, 0 = false, S, T, L, D = device dependent coding.

Class: AD = address, DD = device dependent data, UC = universal command, ST = status data

Application of IEEE Std 488 may be better understood by examining a set of instrument communications, typical remote (carried on the bus) messages and their use in operational sequences. Table 1 lists a few of the 47 remote messages defined by IEEE Std 488. The coding scheme for these messages must be universal and is based on the ASCII code (that is, ANSI X3.4-1977, American National Standard Code for Information Interchange). Devices must be able to generate, process, and receive interface messages with these codes to be compatible with IEEE Std 488. Device-dependent messages (for example, DAB), however, may be sent with any code set although the ASCII code is preferred and used frequently.

The task of a voltmeter measurement further illustrates the interface bus in action. A series of remote messages are sent in the general form UNL, MLA, MTA, DAB1, DAB2, ... , DABn. Stated differently, the controller sends out the unlisten command followed by the voltmeter's listen address (MLA) and the controller's talk address (MTA) with the ATN message true. This is followed by a series of device-dependent program codes with the ATN message sent false. Similarly, the voltmeter may be selected to output its measurement results (for example, DC + 12001E - 03,) to another device with the sequence UNL,MTA,MLA,DAB1,. . . , DABn. In each case, the MTA and MLA messages must be processed by interface related hardware (or software) to establish the proper data paths internal to each device. The associated instrument functionality may be implemented by discrete logic, an LSI chip, state machine designs, or microprocessor and associated memory, for example. Once the proper internal data paths are established, the DAB messages are sent directly from the source (for example, voltage measurement display buffers) within one device to the destination (for example, printer character generator) in another device.

A somewhat more complex situation arises during serial poll sequences. Assume a timer/counter is sending a continuous series of frequency measurements when the time base reference ceases to operate normally. The counter sends an SRQ message and, at some future time, the controller interrupts the measurement process with the remote message sequence: UNL, serial poll enable (SPE), MTA1, MTA2, ... Device one, which may or may not be the counter, then responds with a status byte containing a requested service (RQS) message and possibly an STB message. The process of sending an MTAn message and receiving the corresponding RQSn message is repeated until an affirmative RQS message is received whereupon the appropriate corrective action is taken (for example, reprogram or cease to use the abnormal device). During the serial poll process, both the controller and devices sending status messages must generate or interpret a set of codes to convey the proper status data. These examples serve to illustrate the essential mechanisms of communication used throughout IEEE Std 488.

IEEE Std 728: A Companion Document

Shortly after IEEE Std 488 was published it became obvious that standardization of more than just the communication protocol at the hardware level was needed to facilitate the assembly and use of instrumentation systems. The same committees responsible for the development of IEEE Std 488 and IEC Publication 625 began to formalize possible approaches toward establishing common device dependent message structures. During the early stages, the focus was on two broad message types: (1) program data messages aimed at establishing the internal measurement configuration of a programmable instrument, and (2) measurement data messages aimed at structuring the way in which measurement results are to be sent to other system devices. Eventually more generic types of data structures were incorporated, such as block and string data. The overall result of these efforts is a set of diagrams that establishes the syntactical rules for structuring device dependent messages with very little semantic content. Again, the attempt was made to minimize, if not eliminate, the degree of instrument and system dependence. IEEE Std 728 is, in fact, a Recommended Practice for the reason that in a number of cases alternative means are specified to achieve similar message constructs. The rationale for this approach stems from the fact that the scope and diversity of application for IEEE Std 488 products have grown to the extent that a single fixed syntax is not practical. While IEEE Std 728 is neither as definitive nor extensive as some applications might prefer, it does go a long way toward facilitating the more software oriented aspects of instrumentation system communications.

Global Perspectives

It is important to recognize that the IEEE standards contained in this printing and their international counterparts were elaborated by an international set of participants. Those nations that contributed substantially in the close and harmonious cooperate effort were; France, Germany, Hungary, Japan, The Netherlands, United Kingdom, and United States. While it is not feasible to acknowledge all the individuals concerned, the widespread use of these standards is in the large measure due to their collective and considerable efforts.

When the IEEE Std 488 was introduced in the mid-70's the preponderance of new products utilizing the concepts were either measurement or stimulus instrumentation as might be expected. In early 1978 there were some 300 products (that is, different model numbers) that had been introduced on a global basis. At this time a pattern began to emerge that has remained essentially the same with respect to the types of different products utilizing IEEE Std 488. Approximately 50% of the products are measurement instruments (for example, voltmeters, counters, power meters, network analyzers, spectrum analyzers) and another 20% are in the stimulus category (for example, synthesizers, sweepers, function and waveform generators, power supplies). A broad range of controllers, calculators, minicomputers, and more recently PC's accounts for another 15%. The remaining 15% of the products are a mixture, storage devices (for example, magnetic tapes and discs), display devices (for example, printers, plotters), and special products such as IEEE Std 488 adaptors, bus extenders, environmental chambers and mechanical arms. Today there are in excess of 3000 different model number products supplied by more than 300 manufacturers utilizing IEEE Std 488 or IEC Publication 625-1 capabilities. The advances in microprocessors and LSI technology have facilitated to a large extent the ease with which IEEE Std 488 capability has been incorporated in all types of products.

References

[1] RICCI, D. W. and NELSON, G. E. Standard Instrument Interface Simplifies System Design, *Electronics*, pp 95-106, No 14, 1974.

[2] KNOBLOCK, D. E., LOUGHRY, D. C., and VISSERS, C. A. Insight Into Interfacing, *IEEE Spectrum*, pp 50-57, May 1975.

ANSI/IEEE
Std 488-1978
(Revision of ANSI/IEEE
Std 488-1975)
(Includes supplement
IEEE Std 488A-1980)

An American National Standard

IEEE Standard Digital Interface for Programmable Instrumentation

Sponsor

IEEE Instrumentation and Measurements Society

Approved June 9, 1978

IEEE Standards Board

Approved July 18, 1979

American National Standards Institute

The Institute of Electrical and Electronics Engineers, Inc
345 East 47th Street, New York, N.Y. 10017

NOTE
RELATION TO IEC STANDARDS

(This note is not a part of IEEE Std 488-1978)

IEEE Standard 488-1978 is in concert with its international counterpart, International Electrotechnical Commission (IEC) Publication 625-1*. Major portions of the forthcoming IEC publication and this standard are identical, word for word. A number of paragraphs and clauses in the IEC publication have, for international reasons, received slightly different editorial treatment although the technical content and intended meaning are also identical with this standard. In some instances, there are apparent differences although, in fact, the technical content is identical (for example, the IEC publication assumes, via special notation, whereas this standard explicitly identifies state change exit conditions for IFC). The two standards are, in practice, identical in their functional and electrical specifications. The IEC document uses a mechanical connection system identical to the EIA RS 232-C data communication interface. Component damage, due to incompatible voltage levels, is possible if data communication and instrumentation interfaces are inadvertently interconnected. IEEE and US industry practice is to minimize component damage to programmable instrumentation systems by specifying the connector system of this standard. Any mechanical specification difference between the IEEE and IEC standards may be accommodated by a simple (physical) adapter assembly when products implemented from the two standards are interconnected.

---

*Publication dated 1979.

Fifth Printing

Corrected edition August 26, 1983

Includes supplement IEEE Std 488A-1980

Approved September 27, 1979

IEEE Standards Board

NOTE: The text changes associated with supplement IEEE Std 488A-1980 are identified by a double vertical line in the margin opposite the relevant subject matter. See also page 84 for further clarification.

Foreword

(This Foreword is not a part of ANSI/IEEE Std 488-1978, IEEE Standard Digital Interface for Programmable Instrumentation.)

IEEE Std 488-1975 has enjoyed wide use since its initial publication. Experience with the standard indicates that some portions of the written text could be more clearly stated. A draft international standard IEC Publication 625* using the same interface concepts has been further elaborated since the original issuance of IEEE Std 488 and provides additional points of clarification. The principle improvements to the written text of this revised standard focus on editorial changes rather than substantive technical changes to the standard itself. Further, where changes of a technical nature have been made, they provide clarification of potentially ambiguous statements and broadening of certain specifications. However, the changes do not contradict the technical concepts as defined in the original published standard. Those changes that could have a bearing on the technical compatibility between independently designed products are identified as follows:

- Clauses 2.3.3, 2.4.2, (SH Function), END Message, Table 3, Table 6
- Clause 2.4.2, AH Function, Optional Transition Term, Fig 4
- Clause 2.5.2, TE and T Function, and L Alternate Transition, Fig 6 Note
- Clause 2.5.2, (TE Message Outputs), END Message, Table 9
- Clause 2.6.2, LE and T Function, Alternate Transition, Fig 8, Table 17 Notes
- Clause 2.8.1, RL and LE Function, Alternate Transition, Fig 10
- Clause 2.12.2, C Function, Transitions Clarified, Fig 14
- Clause 2.13.3, Valid Uniline Message Transfer Times
- Clause 2.13.5, Remote Message Coding, Perspectives, Table 38
- Clause 3.3.2, Driver Specifications, Schottky Levels
- Clause 3.4.2, Receiver Specification, Preferred
- Clause 3.7.2, 4.3, Cable Construction
- Clause 4.4, Connector Mounting, Dimensional Tolerances, Fig 18
- Clause 5.2.3 Data Rate Consideration, Higher Speed Operation
- Clause 6.2.4, Devices Powered Off
- Clause 6.3.1, UNT use
- Clause 6.5, Operational Sequences, Exit Condition Caution
- Clause 6.5.2, Serial Poll Completion
- Clause 6.5.4, Parallel Poll Configure, Recommended Sequence
- Appendix C, Interface Function Capability Identification

NOTE: Text changes associated with the above mentioned clauses are identified by a vertical line in the margin opposite the relevant subject matter.

The IEEE Std 488-1978, IEEE Standard Digital Interface for Programmable Instrumentation, deals with systems that use a byte-serial bit-parallel means to transfer digital data among a group of instruments and system components. The interface system described herein is optimized as an interdevice interface for system components in relatively close proximity able to communicate over a contiguous party-line bus system.

This document contains seven sections as follows.

Section 1 contains the scope, the object, basic definitions, and summary description of the interface.

Section 2 deals with functional concepts and specifications of the interface system described in this standard. One or more interface functions contained within a device are each able to process messages and change states to maintain an orderly flow of information among a set of inter-connected devices.

Section 3 deals with the electrical realization of the interface in order to transfer messages among a set of interconnected devices.

Section 4 deals with the mechanical realization of the interface in order to implement the electrical aspects of the interface system.

---

*Will be published as IEC Publication 625-1.

Section 5 deals with system considerations that must be given to the design of an individual device in order to make it compatible with other devices of a measurement system.

Section 6 deals with system considerations that must be recognized by the user of devices designed in accordance with this standard.

Appendixes deal with explanatory matter and examples.

In order to design equipment in accordance with this standard, the designer must have read the entire document.

In order to interconnect and program equipment designed in accordance with this standard, the user should have knowledge of Sections 1 and 6. If the coding and transfer of messages is not done automatically by the apparatus to be programmed, it will be necessary that the user have knowledge of Section 2. General familiarity with the other sections is recommended. The user must also be familiar with device-dependent characteristics of apparatus that may be used in a system, but that are beyond the scope of this standard.

This standard defines an interface with the objective to assure that messages may be accurately communicated between two or more devices in a system, but does not guarantee that each device will interpret properly all possible messages sent to it or will properly generate all necessary messages. A wide latitude of interface capability is permitted within the scope of this standard which may permit operational incompatibility among interconnected devices.

A device designer must have sufficient awareness of the characteristics of systems which might include his device in order to select correctly among the options provided in this standard. Likewise, a system configurator must have sufficient awareness of the options included in each of the devices in his system in order to ensure that the correct communication techniques are used.

This standard does not specify the device-dependent or operational characteristics required for complete system compatibility. Therefore, following the rules and procedures of this standard alone will not guarantee unconditional compatibility.

The interface specified by this standard includes patented matter which pertains only to the three-wire handshake as defined throughout 2.3 and 2.4 and summarized in Appendix B.

The IEEE Standards Office calls attention to the fact that it is claimed that the three-wire handshake referenced in 2.3 and 2.4 is the subject of one US patent and corresponding patents of foreign countries owned by the Hewlett-Packard Company. Although these patents appear to cover the subject of 2.3 and 2.4 in this standard, the IEEE takes no position with respect to patent validity. The Hewlett-Packard Company has assured the IEEE that it is willing to grant a license under these patents on reasonable and nondiscriminatory terms and conditions to anyone wishing to obtain such a license. The Hewlett-Packard Company's undertakings in this respect are on file with the IEEE Standards Office and the license details may be obtained from the legal department of the Hewlett-Packard Company whose address is 1501 Page Mill Road, Palo Alto, Calif 94304.

This standard is based on work initiated by the International Electrotechnical Commission (IEC) within Technical Committee 66, Working Group 3, and follows the general concepts of a standard prepared by the IEC. This standard influenced, and was influenced by, working documents in the IEC.

The two IEEE Committees responsible for the preparation and evaluation of this standard within the US were the Instrumentation and Measurements Group Subcommittee on Instrument/Computer Interfaces (which also served as the US Advisory Committee to US representatives on IEC TC 66/WG3) and the Instrumentation and Measurements Group Technical Committee on Automated Instrumentation.

Personnel for the Instrument/Computer Interfaces Subcommittee were:

R. G. Fulks, *Chairman* | **D. C. Loughry**, *Secretary*

L. Costrell
H. Draye
G. Foley
M. Gipe
M. Graube
T. J. Harrison
D. Lehmer
J. Pepin
B. Perlman
D. Picker
L. C. Sollman

Personnel for the Automated Instrumentation Technical Committee were:

F. Liguori, *Chairman* **C. W. Osborn**, *Secretary*

O. R. Batchelder
B. Bendel
R. D. Chew
D. M. Goodman
J. A. Houston
R. P. Keats
D. C. Loughry
R. H. Nesson
J. K. Skilling
M. G. Slade
P. Sylvan

When this standard was approved June 9, 1978, the IEEE Standards Board had the following membership:

Joseph L. Koepfinger, *Chairman* **Irvin N. Howell, Jr**, *Vice Chairman*

Ivan G. Easton, *Secretary*

William E. Andrus
C. N. Berglund
Edward J. Cohen
Warren H. Cook
David B. Dobson
R. O. Duncan
Charles W. Flint
Jay Forster
Ralph I. Hauser
Loering M. Johnson
Irving Kolodny
William R. Kruesi
Thomas J. Martin
John E. May
Donald T. Michael
Voss A. Moore
William S. Morgan
Robert L. Pritchard
Blair A. Rowley
Ralph M. Showers
B. W. Whittington

IEEE Standards documents are developed within the Technical Committees of the IEEE Societies and the Standards Coordinating Committees of the IEEE Standards Board. Members of the committees serve voluntarily and without compensation. They are not necessarily members of the Institute. The standards developed within IEEE represent a consensus of the broad expertise on the subject within the Institute as well as those activities outside of IEEE which have expressed an interest in participating in the development of the standard.

Use of an IEEE Standard is wholly voluntary. The existence of an IEEE Standard does not imply that there are no other ways to produce, test, measure, purchase, market, or provide other goods and services related to the scope of the IEEE Standard. Furthermore, the viewpoint expressed at the time a standard is approved and issued is subject to change brought about through developments in the state of the art and comments received from users of the standard. Every IEEE Standard is subjected to review at least once every five years for revision or reaffirmation. When a document is more than five years old, and has not been reaffirmed, it is reasonable to conclude that its contents, although still of some value, do not wholly reflect the present state of the art. Users are cautioned to check to determine that they have the latest edition of any IEEE Standard.

Comments for revision of IEEE Standards are welcome from any interested party, regardless of membership affiliation with IEEE. Suggestions for changes in documents should be in the form of a proposed change of text, together with appropriate supporting comments.

Interpretations: Occasionally questions may arise regarding the meaning of portions of standards as they relate to specific applications. When the need for interpretations is brought to the attention of IEEE, the Institute will initiate action to prepare appropriate responses. Since IEEE Standards represent a consensus of all concerned interests, it is important to ensure that any interpretation has also received the concurrence of a balance of interests. For this reason IEEE and the members of its technical committees are not able to provide an instant response to interpretation requests except in those cases where the matter has previously received formal consideration.

Comments on standards and requests for interpretations should be addressed to:

Secretary, IEEE Standards Board
345 East 47th Street
New York, NY 10017
USA

Contents

FIGURES PAGE

TABLES

A HELPFUL NOTE TO USERS OF IEEE STD 488-1978

(This note is not a part of IEEE Std 488-1978)

IEEE Std 488-1978, Digital Interface for Programmable Instrumentation, specifies mechanical threaded fasteners per metric thread size ISO M3.5X0.6 as defined throughout Section 4, Mechanical Specifications.

The IEEE Standards Office calls your attention to the fact that a number of products designed for use with IEEE Std 488-1978 contain nonmetric threaded fasteners in contradiction to 4.4 (Fig 18) and 4.5 (Fig 20) of the standard. This situation results from the fact that some products otherwise in conformity with the standard were designed and manufactured before IEEE Std 488-1975 was approved and published. Most suppliers of instrumentation products now utilize metric threaded fasteners and fully satisfy IEEE Std 488-1978.

A number of manufacturers now use metric threaded fasteners with a black finish (for example, copper sulfate over nickel) to designate the use of metric threads whereas the older nonmetric parts have a shiny nickel finish. In addition, these parts are stamped with a small letter M on the top surface to further indicate the metric part.

In the interest of minimizing mechanical incompatibility problems for both manufacturers and users of this standard, it is recommended that the metric parts defined by 4.4 and 4.5 of IEEE Std 488-1978 be identified with a black finish to indicate and emphasize the use of metric threads.

Further details may be obtained from the IEEE Standards Office as to the technical specifications and typical sources of supply for metric threaded parts. Cooperation in this matter will facilitate the implementation and use of IEEE Std 488-1978.

An American National Standard

IEEE Standard Digital Interface for Programmable Instrumentation

1. General

1.1 Scope. This standard applies to interface systems used to interconnect both programmable and nonprogrammable electronic measuring apparatus with other apparatus and accessories necessary to assemble instrumentation systems.

It applies to the interface of instrumentation systems, or portions of them, in which the:

(1) Data exchanged among the interconnected apparatus is digital (as distinct from analog)

(2) Number of devices that may be interconnected by one contiguous bus does not exceed 15

(3) Total transmission path lengths over the interconnecting cables does not exceed 20 m

(4) Data rate across the interface on any signal line does not exceed 1 Mb/s.

The basic functional specifications of this standard may be used in digital interface applications which require longer distances, more devices, increased noise immunity, or combinations of these. Different electrical and mechanical specifications may be required (for example, symmetrical circuit configurations, high threshold logic, special connectors, or cable configurations) for these extended applications.

This standard may also be applicable to other instrumentation system elements such as processors, stimulus, display, or storage devices, and terminal units found useful in instrumentation systems.

It applies generally to laboratory and production test environments which are both electrically quiet and restricted as to physical dimensions (distances between the system components).

This standard deals only with the interface characteristics of instrumentation systems to the exclusion of design specifications, consideration of radio-interface regulations, performance requirements, and safety requirements of apparatus.

NOTE: For the latter two items, reference is made to IEC-Publ 348 (1971), Safety Requirements for Electronic Measuring Apparatus, and IEC-Publ 359 (1971), Expression of the Functional Performance of Electronic Measuring Equipment.[1]

Throughout this standard, and insofar as further distinction is not necessary, the term "system" denotes the byte-serial bit-parallel interface system that, in general, includes all circuits, cables, connectors, message repertoire, and control protocol to effect unambiguous data transfer between devices; and the term "device" or "apparatus" denotes any programmable measurement device or other product connected to the interface system that communicates information via, and conforms to, the interface system definition.

A primary focus of this standard is to set forth an interface system to interconnect self-contained apparatus to other apparatus by external means. This same standard may be applied to interconnecting the internal subsections within a self-contained equipment.

1.2 Object. This standard is intended:

(1) To define a general-purpose system for use in limited distance applications

(2) To specify the device-independent mechanical, electrical, and functional interface requirements that the apparatus shall meet in order to be interconnected and communicate unambiguously via the system.

[1]IEC documents are available from the American National Standards Institute, 1430 Broadway, New York, NY 10018.

(3) To specify the terminology and definitions related to the system

(4) To enable the interconnection of independently manufactured apparatus into a single functional system

(5) To permit apparatus with a wide range of capability — from the simple to the complex — to be interconnected to the system simultaneously

(6) To permit direct communication between the apparatus, without requiring all messages to be routed to a control or intermediate unit

(7) To define a system with a minimum of restrictions on the performance characteristics of the apparatus connected to the system

(8) To define a system that permits asynchronous communication over a wide range of data rates

(9) To define a system that, of itself, may be relatively low cost and permits the interconnection of low cost devices

(10) To define a system that is easy to use.

1.3 Definitions. The following definitions apply for the purpose of this standard.

This section contains only general definitions. Detailed definitions are given in further sections, as appropriate.

1.3.1 General System Terms

compatibility. The degree to which devices may be interconnected and used, without modification, when designed as defined throughout this standard (for example, mechanical, electrical, or functional).

handshake cycle. The process whereby digital signals effect the transfer of each data byte across the interface by means of an interlocked sequence of status and control signals. Interlocked denotes a fixed sequence of events in which one event in the sequence must occur before the next event may occur.

interface. A common boundary between a considered system and another system, or between parts of a system, through which information is conveyed.

interface system. The device-independent mechanical, electrical, and functional elements of an interface necessary to effect communication among a set of devices. Cables, connector, driver and receiver circuits, signal line descriptions, timing and control conventions, and functional logic circuits are typical interface system elements.

local control. A method whereby a device is programmable by means of its local (front or rear panel) controls in order to enable the device to perform different tasks. (Also referred to as manual control.)

programmable. That characteristic of a device that makes it capable of accepting data to alter the state of its internal circuitry to perform a specific task(s).

remote control. A method whereby a device is programmable via its electrical interface connection in order to enable the device to perform different tasks.

system. A set of interconnected elements constituted to achieve a given objective by performing a specified function.

1.3.2 Units Connected Via the Interface System

programmable measuring apparatus. A measuring apparatus that performs specified operations on command from the system and, may transmit the results of the measurement(s) to the system.

terminal unit. An apparatus that terminates the considered interface system and by means of which a connection (and translation, if required) is made between the considered interface system and another external interface system.

1.3.3 Signals and Paths

bidirectional bus. A bus used by any individual device for two-way transmission of messages, that is, both input and output.

bit-parallel. Refers to a set of concurrent data bits present on a like number of signal lines used to carry information. Bit-parallel data bits may be acted upon concurrently as a group (byte) or independently as individual data bits.

bus. A signal line or a set of signal lines used by an interface system to which a number of devices are connected and over which messages are carried.

byte. A group of adjacent binary digits operated on as a unit and usually shorter than a computer word (frequently connotes a group of eight bits).

byte-serial. A sequence of bit-parallel data bytes used to carry information over a common bus.

high state. The relatively more positive signal level used to assert a specific message content associated with one of two binary logic states.

low state. The relatively less positive signal level used to assert a specific message content associated with one of two binary logic states.

signal. The physical representation of information.

NOTE: For the purpose of this standard, this is a restricted definition of what is often called "signal" in more general terms, and is hereinafter referred to digital electrical signals only.

signal level. The magnitude of signal compared to an arbitrary reference magnitude (voltage in the case of this standard).

signal line. One of a set of signal conductors in an interface system used to transfer messages among interconnected devices.

signal parameter. That parameter of an electrical quantity whose values or sequence of values convey information.

unidirectional bus. A bus used by any individual device for one-way transmission of messages only, that is, either input only or output only.

1.4 Interface System Overview

1.4.1 Interface System Objective. The overall purpose of an interface system is to provide an effective communication link over which messages are carried in an unambiguous way among a group of interconnected devices.

Messages (quantities of information) carried by an interface system belong to either of two broad categories:

(1) Messages used to manage the interface system itself, hereinafter called interface messages

(2) Messages used by the devices interconnected via the interface system that are carried by, but not used or processed by the interface system directly, hereinafter called device dependent messages

NOTE: The detailed specification of device dependent messages is beyond the scope of this standard.

1.4.2 Fundamental Communication Capabilities. An effective communication link requires three basic functional elements to organize and manage the flow of information to be exchanged among devices:

(1) A device acting as a listener
(2) A device acting as a talker
(3) A device acting as a controller

In the context of the interface system described by this standard:

(1) A device with the capability to listen can be addressed by an interface message to receive device dependent messages from another device connected to the interface system.

(2) A device with the capability to talk can be addressed by an interface message to send device dependent messages to another device connected to the interface system.

(3) A device with the capability to control can address other devices to listen or to talk. In addition, this device can send interface messages to command specified actions within other devices. A device with only this capability neither sends nor receives device dependent messages.

NOTE: The use of the word controller throughout this standard applies strictly to the management (control) of the interface system and does not imply the broad capabilities typically associated with the word in the data processing context. Further classification of the controller will be made in Section 2 to distinguish between different types of controller capabilities related to the interface system.

Listener, talker, and controller capabilities occur individually or in any combination in devices interconnected via the interface system as shown in Fig 1.

In addition to the basic listener, talker, and controller functions, the system provides interface messages to accomplish the following operations:

(1) A serial poll sequence may be initiated when a device (with talker function) requires some action by the controller, by transmitting the service request message. The controller will then obtain the status byte of all possible devices in sequence to ascertain which required service.

(2) The Parallel Poll function provides a device with the ability to transmit on the controller's demand one bit of status information (request service) simultaneously with several other devices. The assignment of a data line to a particular device for the response to a parallel poll may be accomplished through interface messages.

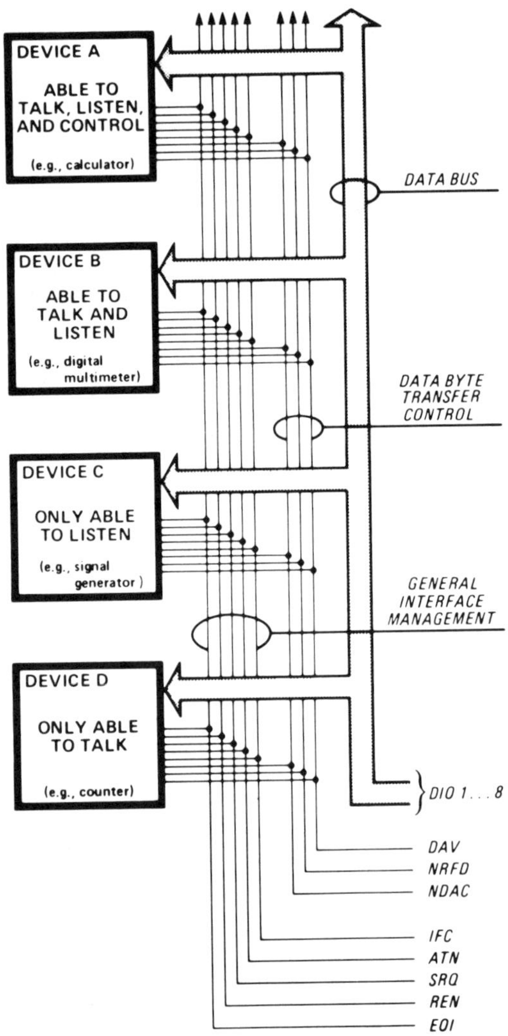

Fig 1
Interface Capabilities and Bus Structure

(3) The Device Clear and Device Trigger functions provide a device with the ability to be initialized or triggered, on command from the controller. This may occur simultaneously with other selected or all devices in a system.

(4) The remote/local function provides a device with the ability to select between program data from the bus and local data (for example, front panel controls.)

1.4.3 Message Paths and Bus Structure. The interface system contains a set of sixteen signal lines used to carry all information, interface messages, and device dependent messages among interconnected devices.

Messages may be coded on one or a set of signal lines as determined by the particular message content and its relationship to the interface system.

The bus structure is organized into three sets of signal lines:

(1) Data bus, 8 signal lines

(2) Data byte transfer control bus, 3 signal lines

(3) General interface management bus, 5 signal lines

Fig 1 illustrates the basic communication paths.

A set of eight interface signal lines carries all 7 bit interface messages and the device dependent messages:

(1) DIO1 (data input output 1)

. .

. .

. .

(8) DIO8 (data input output 8)

Message bytes are carried on the DIO signal lines in a bit-parallel byte-serial form, asynchronously, and generally in a bidirectional manner.

NOTE: A message may be carried on an individual DIO signal line when required.

A set of three interface signal lines is used to effect the transfer of each byte of data on the DIO signal lines from a talker or controller to one or more listeners:

(1) DAV (data valid) is used to indicate the condition (availability and validity) of information on the DIO signal lines

(2) NRFD (not ready for data) is used to indicate the condition of readiness of device(s) to accept data

(3) NDAC (not data accepted) is used to indicate the condition of acceptance of data by device(s)

The DAV, NRFD, and NDAC signal lines operate in what is called a three-wire (interlocked) handshake process to transfer each data byte across the interface.

Five interface signal lines are used to manage an orderly flow of information across the interface:

(1) ATN (attention) is used (by a controller) to specify how data on the DIO signal lines

are to be interpreted and which devices must respond to the data

(2) IFC (interface clear) is used (by a controller) to place the interface system, portions of which are contained in all interconnected devices, in a known quiescent state

(3) SRQ (service request) is used by a device to indicate the need for attention and to request an interruption of the current sequence of events

(4) REN (remote enable) is used (by a controller) in conjunction with other messages to select between two alternate sources of device programming data

(5) EOI (end or identify) is used (by a talker) to indicate the end of a multiple byte transfer sequence or, in conjunction with ATN (by a controller), to execute a polling sequence

1.4.4 Interface System Elements. The primary elements of this interface system are:

(1) Functional elements
(2) Electrical elements
(3) Mechanical elements

Each is described in a following section.

2. Functional Specifications

2.1 Functional Partition. A device is a physical entity designed for a particular application. It may be partitioned conceptually into three major functional areas each containing unique capabilities:

(1) Device functions (definition is application dependent)

(2) Interface functions (definition is application independent)

(3) Message coding logic

All communication to or from interface functions is defined in terms of messages and state linkages (see 2.1.3).

All messages carried on the signal lines are coded according to the coding logic defined in 2.13.

2.1.1 Device Functions. The scope, purpose, size, content, and organization of the device function area (for example, analog signal measurement capability, range, modes of operation, etc) are beyond the scope of this standard. Fig 2 illustrates the device function area *B* for which the designer has complete freedom to define device related capability and the interface function area *A* for which the designer has no freedom to define new capability beyond that specified in this standard.

2.1.2 Interface Function Concepts

2.1.2.1 *Interface Functions.* An interface function is the system element which provides the basic operational facility through which a device can receive, process, and send messages. A number of interface functions, each of which acts in accordance with specific protocol, are defined throughout this section of the standard. Each specific interface function may only send or receive a limited set of messages within particular classes of messages.

2.1.2.2 *Interface Function State.* Each of the interface functions is defined in terms of one or more groups of interconnected, mutually exclusive states.

One and only one state shall be active at any one time within a single group of interconnected, mutually exclusive states.

For each state of an interface function definitions are given for the following:

(1) Messages that may or must be sent over the interface while that state is active

(2) Conditions under which the function must leave that state and enter one of the other states in its group

These messages and conditions define the processing capability of the state.

2.1.2.3 *Interface Function Repertoire.* The designer is given the choice to select the particular set of interface functions necessary to fit the particular device application area. Fig 2 and Table 1 identify the available interface functions.

The total processing capability of a set of interface functions (designer selected set included in a specific device) at any moment is the logic conjunction of the processing capabilities of all those states (within each individual interface function) that are active at that moment.

2.1.2.4 *Interface Function Assumptions and Perspective.* The state diagrams used to define the interface functions do not indicate, either explicitly or implicitly, the intended existence of specific circuit elements to achieve the logical and physical implementation of a function. For example, not all states necessarily imply the existence of a latched flip-flop or other memory element.

The state diagrams used to define the interface functions are intended to permit the use of a wide variety of logic circuit implementations (for example, random logic, sequential logic, etc).

The designer is free to combine and implement two or more interface functions with one logic design provided all the conditions for each state of each interface function as defined in this section are met.

Throughout this section of the standard, the state diagrams, written descriptions, requirements, and guidelines are written for and should be interpreted from the device perspective. Sections 5 and 6 will describe the interaction among devices from the system perspective.

An interface function must ignore (not respond to) any message coding not specifically defined.

A function may stay in any state for any amount of time (including zero) after exit conditions are met if this is not in conflict with specified constraints.

2.1.3 Message Concepts

2.1.3.1 *Message.* Each message represents a quantity of information and will be received either true or false at any specific time.

All communication between an interface function and its environment is accomplished through messages sent or received.

Fig 2
Functional Partition Within a Device

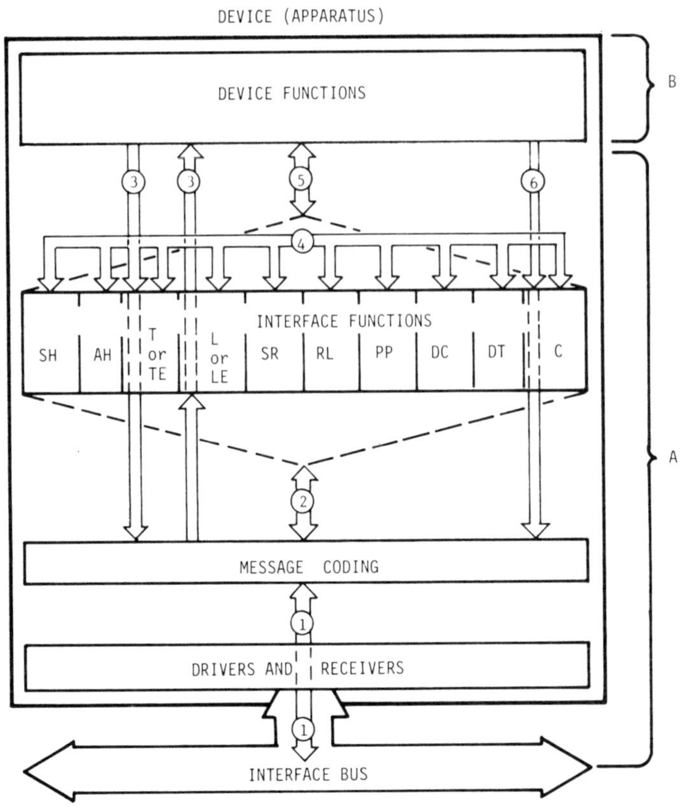

A = Capability defined by this standard
B = Capability defined by the designer
1 = Interface bus signal lines
2 = Remote interface messages to and from interface functions
3 = Device dependent messages to and from device functions
4 = State linkages between interface functions
5 = Local messages between device functions and interface functions (messages to interface functions are defined, messages from interface functions exist according to the designer's choice)
6 = Remote interface messages sent by device functions within a controller

Table 1
Interface Function Repetoire

Interface Function	Symbol	Relevant Message Paths
Source handshake	SH	1, 2, 4, 5
Acceptor handshake	AH	1, 2, 4, 5
Talker or extended talker	T or TE	1, 2, 3, 4, 5
Listener or extended listner	L or LE	1, 2, 3, 4, 5
Service request	SR	1, 2, 4, 5
Remote local	RL	1, 2, 4, 5
Parallel poll	PP	1, 2, 4, 5
Device clear	DC	1, 2, 4, 5
Device trigger	DT	1, 2, 4, 5
Controller	C	1, 2, 4, 5, 6

2.1.3.2 *Local Message Route and Content.* Messages sent between a device function and an interface function are called local messages.

Local messages flow between device functions and interface functions; see Fig 2, message route 5.

NOTE: Certain local messages are conveyed as remote messages and vice versa.

The designer is not allowed to introduce new local messages to interface functions.

The designer is allowed to introduce a local message derived from any state of any interface function to device function(s).

Local messages sent by device functions must exist for enough time to cause the required state transitions.

2.1.3.3 *Remote Message Route and Content.* Messages sent via the interface between interface functions of different devices are called remote messages.

Each remote message is either an interface message or a device dependent message.

Each interface message is sent to cause a state transition within another interface function. An interface message will not be passed along to the device when received by an interface function as shown in Fig 2, message route 2.

Device dependent messages are passed between the device functions and the message coding logic via specified interface functions. These will cause no state transitions within the interface functions. Examples of device dependent messages include device programming data, device measurement data, and device status data as shown in Fig 2, message route 3.

2.1.3.4 *State Linkage Route and Content.* A state linkage is the logical interconnection of two interface functions where the transition to an active state of one interface function is dependent on the existence of a specified active state of another interface function as indicated in Fig 2, message route 4.

2.1.3.5 *Message Coding.* Message coding is the act of translating remote messages to or from interface signal line values.

A message sent over a single line is called a uniline message. Two or more of these messages can be sent concurrently.

A message that shares a group of signal lines with other messages, in some mutually exclusive set, is called a multiline message. Only one multiline message (message byte) can be sent at one time.

2.1.3.6 *Classification of Multiline Messages.* Multiline messages are interpreted as interface messages when the ATN message is true.

Multiline messages are interpreted as device dependent messages when the ATN message is false.

Each device, when the ATN message is true, may accept and process any or all of these specific classes of multiline messages as determined both by the extent of the interface function capability contained within the device and the current set of active interface function states:

(1) Universal commands (all devices)

(2) Addressed commands (all devices addressed to listen)

(3) Addresses (all devices)

(4) Secondary addresses or commands (all devices enabled by a primary address or command)

For a list of specific multiline commands see Table 36.

2.1.3.7 *Message Transfer Conventions*

2.1.3.7.1 *Remote Message Transfer Conventions*

(1) The value (true or false) of all remote messages capable of being sent by a device shall at all times be as dictated by active states of its interface functions.

(2) The interface signal line(s) used to send a message value shall be set to the levels specified by Table 38, Remote Message Coding.

(3) Since normal interface operation allows two devices to simultaneously send opposite values of the remote messages, a technique shall be provided for resolving these conflicts. This is accomplished by implementing two types of message transfer over the interface, active transfer and passive transfer. The interface is structured so that in all conflicts between two message values, one of them will be active and the other passive. Messages shall be transferred in such a way that the active value overrides the passive value in every conflict that arises.

(4) A remote message can be transferred in one of four ways:

(a) An active true value being sent is guaranteed to be the value received, and the device need not allow it to be overridden

(b) A passive true value being sent is not guaranteed to be the value received, and the device must allow it to be overridden

(c) An active false value being sent is guaranteed to be the value received, and the device need not allow it to be overridden

(d) A passive false value being sent is not guaranteed to be the value received and the device must allow it to be overridden

(5) Throughout the text, the terms true and false if not qualified are assumed to mean active true and active false during all discussions of remote message values sent by an interface function.

(6) For two specific remote messages, DAC and RFD, only false values are defined to be sent actively. Thus, an AND operation can be considered to be performed on the interface signal lines (see 5.4).

(7) For one remote message, SRQ, only true values are defined to be sent actively. Thus, an OR operation can be considered to be performed on the interface signal lines (see 5.4).

(8) Only the multiline message(s) to be sent true will be specified for an interface function state since multiline messages (sent via the DIO lines) are by their nature mutually exclusive. It should be understood that all unspecified multiline messages are sent passive false while the state is active.

2.1.3.7.2 *Local Message Transfer Conventions*

(1) The coding of local messages is beyond the scope of this standard and is left to the discretion of the device designer.

(2) It is recommended that local messages qualifying transitions within any group of mutually exclusive states of an interface function be themselves mutually exclusive.

2.2 Notation Used to Specify Interface Functions

2.2.1 State Diagram Notation. Each state that an interface function can assume is represented graphically as a circle. A four-letter upper-case mnemonic, always ending in an S, is used within the circle to identify the state:

All permissible transitions between states of an interface function are represented graphically by arrows between them.

Each transition is qualified by an expression whose value is either true or false. The interface function shall remain in its current state if all expressions which qualify transitions leading to other states are false. The interface function shall enter the state pointed to if, and only if, one of these expressions becomes true. The new state may be entered at any time after the expression(s) become(s) true, unless a time value is specified.

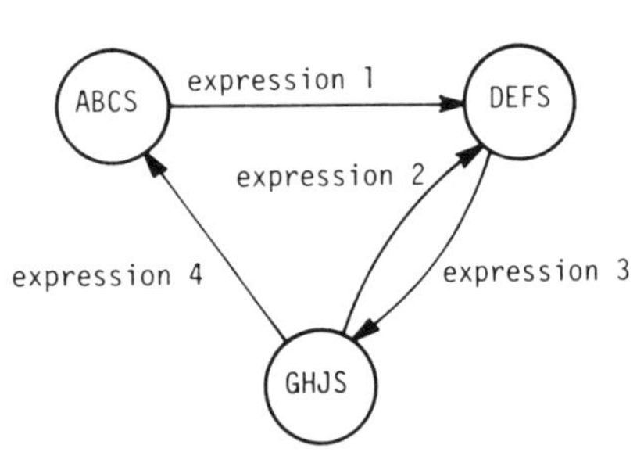

An expression consists of one or more local messages, remote messages, state linkages, or minimum time limits used in conjunction with the operators AND, OR, or NOT.

A local message to an interface function is represented by a three-letter mnemonic written in lower case; for example, rdy.

A remote message (received via the interface) is represented by a three-letter mnemonic written in upper case; for example, ATN.

A linkage from another state diagram is represented by a four-letter mnemonic enclosed in an oval; for example, (LACS). A state linkage is true if the enclosed state is currently active; otherwise, it is false.

A minimum time limit is represented by the symbol T_n. This symbol achieves a true value only after the interface has been in the state originating the corresponding transition for the time value specified. It will remain true until the state is exited. The values for these time limits are contained in Table 39.

The AND operator is represented by the symbol $\wedge$.

The OR operator is represented by the symbol $\vee$.

The AND operator takes precedence over the OR operator within an expression unless otherwise specified by parentheses.

The NOT operator is represented by a horizontal bar placed over the portion of the expression to be negated. The resulting negated expression has a true value if and only if the value of the expression under the bar is false.

If a transition is further qualified by a maximum time limit (within t_n), then the state pointed to shall be entered within the specified amount of time after the expression becomes true. The values for these time limits are contained in Table 39.

If a portion of an expression is optional in that its true value is not required for the complete expression to be true (at the designer's choice), then it is enclosed within square brackets [. . .].

If a specific expression causes a transition to a state from all other states of the diagram, a shorthand notation is used instead of all the individual transitions being drawn. An arrow without a state at its origin is used to represent this condition, and is assumed to originate in all states (for example, IFC or $\overline{\text{REN}}$):

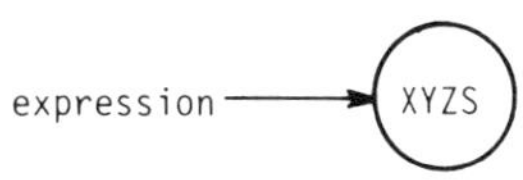

Although POFS (power-off) is a valid state of most interface functions and should normally be shown on all diagrams with a transition leading to the state to be entered at power-on time, a shorthand form is used showing the pon pseudomessage originating a transition to the first state to be entered when power is turned on:

(1) Abbreviated notation used on state diagram:

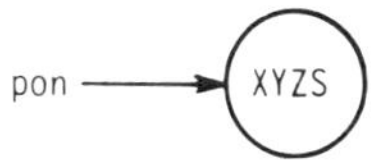

(2) Complete representation implied by preceding symbol:

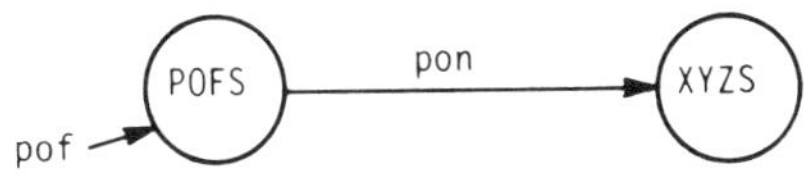

2.2.2 Message Output Notation. The message output table included with each interface function state diagram summarizes only the remote messages allowed to be sent during each of the states of the function.

Rows of the table are used to indicate states of the interface function.

Columns of the table are used to indicate remote messages allowed to be sent during at least one state of the interface function.

Each table entry indicates the value of a message that shall be sent while a specified state is active:

(1) T indicates active true

(2) F indicates active false

(3) (T) indicates passive true

(4) (F) indicates passive false

One column in each table is allocated, if required, to the group of multiline remote messages allowed to be sent. The multiline message to be sent true during each state is placed in its corresponding table entry. False values are not shown since multiline messages are mutually exclusive. Parentheses around a multiline message name specify that it shall be sent passive rather than active true.

A separate column for device function interaction summarizes the corresponding types of messages (or resultant action) device function(s) are allowed to send or receive. Local message(s), beyond the scope of this standard, from the interface function to the device functions may be used to coordinate the appropriate action at the choice of the designer.

2.3 SH (Source Handshake) Interface Function

2.3.1 General Description. The SH interface function provides a device with the capability to guarantee the proper transfer of multiline messages. An interlocked handshake sequence between the SH function and one or more AH (acceptor handshake) functions (each contained within separate devices) guarantees asynchronous transfer of each multiline message. The SH interface function controls the initiation of, and termination of, the transfer of a multiline message byte. This function utilizes the DAV, RFD (ready for data), and DAC (data accepted) messages to effect each message byte transfer.

2.3.2 SH Function State Diagram. The SH function shall be implemented so as to perform according to the state diagram given in Fig 3 and the state descriptions given throughout 2.3. Table 2 specifies the set of messages and states required to effect transition from one active state to another. Table 3 specifies the messages that shall be sent and the device function interaction required while each state is active.

2.3.3 SH Function State Descriptions

2.3.3.1 *SIDS (Source Idle State).* In SIDS the SH interface function is not engaged in the handshake cycle and does not have a new message byte available. The SH function powers on in SIDS.

In SIDS the DAV message shall be sent passive false.

The SH function shall exit SIDS and enter the SGNS (source generate state) if:

(1) The TACS (talker active state) is active

(2) Or the SPAS (serial poll active state) is active

(3) Or the CACS (controller active state) is active

2.3.3.2 *SGNS (Source Generate State).* In SGNS the device is generating a new message byte and the function is waiting for the new byte to become available.

In SGNS the SH function shall send the DAV message false. In this state the device may change the multiline message being sent via the talker or controller interface function while in TACS or CACS.

The SH function shall exit SGNS and enter:

(1) The SDYS (source delay state) if the nba (new byte available) message is true

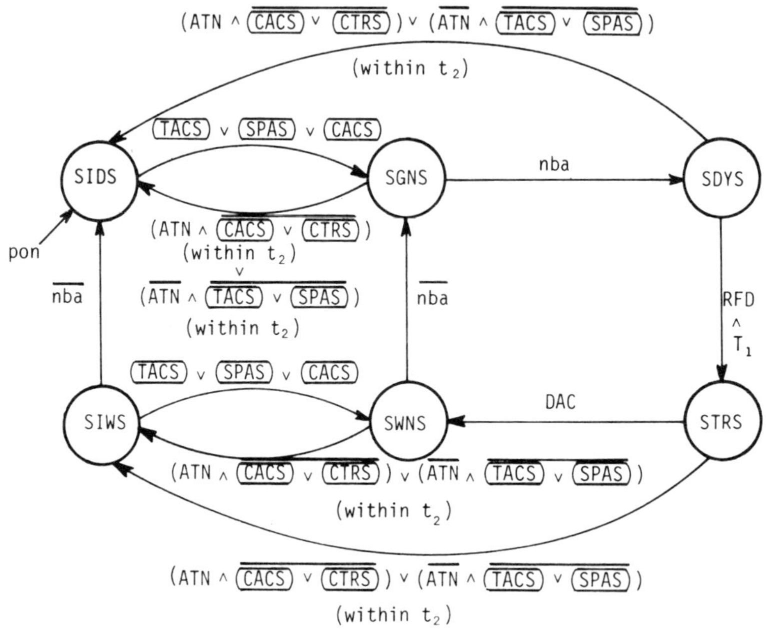

Fig 3
SH State Diagram

(2) The SIDS within t_2 if either:

(a) The ATN message is true and neither CACS nor CTRS is active

(b) Or the ATN message is false and neither TACS nor SPAS is active

2.3.3.3 *SDYS (Source Delay State).* In SDYS the SH function is waiting for a message byte to settle on the interface signal lines after the change during SGNS, and for all the acceptor functions to indicate their readiness to accept the message byte.

In SDYS the SH function shall send the DAV message false. In this state the device shall not change the multiline message being sent.

The SH function shall exit SDYS and enter:

(1) The STRS (source transfer state) only after T_1 if the RFD message is true

(2) The SIDS within t_2 if either:

(a) The ATN message is true and neither CACS nor CTRS is active

(b) Or the ATN message is false and neither TACS nor SPAS is active

2.3.3.4 *STRS (Source Transfer State).* In STRS the SH function indicates to the AH function that it is continuously sending a valid message byte.

In STRS the SH function shall send the DAV message true. In this state the device shall not change either the multiline message or the END message (if used) being sent.

The SH function shall exit STRS and enter:

(1) The SIWS (source idle wait state) within t_2 if either:

(a) The ATN message is true and neither CACS nor CTRS is active

(b) Or the ATN message is false and neither TACS nor SPAS is active.

NOTE: This implies an asynchronous interrupt (see 2.12.5).

(2) The SWNS (source wait for new cycle state) if the DAC message is true

2.3.3.5 *SWNS (Source Wait for New Cycle State).* In SWNS the SH function is waiting for

Table 2
SH Mnemonics

Messages		Interface States	
pon	= power on	SIDS	= source idle state
nba	= new byte available	SGNS	= source generate state
ATN	= attention	SDYS	= source delay state
RFD	= ready for data	STRS	= source transfer state
DAC	= data accepted	SWNS	= source wait for new cycle state
		SIWS	= source idle wait state
		(TACS)	= talker active state (T function)
		(SPAS)	= serial poll active state (T function)
		(CACS)	= controller active state (C function)
		(CTRS)	= controller transfer state (C function)

Table 3
SH Message Outputs

SH State	Remote Message Sent DAV	Device Function (DF) Interaction
SIDS	(F)	DF can change remote multiline messages
SGNS	F	DF can change remote multiline messages
SDYS	F	DAB, EOS multiline, and END messages shall not change
STRS	T	DAB, EOS multiline, and END messages shall not change
SWNS	T or F	DF requested to change multiline messages
SIWS	(F)	DF requested to change multiline messages

Table 4
Allowable Subsets to SH Function

Identification	Description	States Omitted	Other Requirements	Other Function Subsets Required
SH∅	no capability	all	none	none
SH1	complete capability	none	none	T1-T8, TE1-TE8, or C5-C28

Table 5
AH Mnemonics

Messages		Interface States	
pon	= power on	AIDS	= acceptor idle state
rdy	= ready for next message	ANRS	= acceptor not ready state
tcs	= take control synchronously*	ACRS	= acceptor ready state
ATN	= attention	ACDS	= accept data state
DAV	= data valid	AWNS	= acceptor wait for new cycle state
		(LADS)	= listener addressed state (L function)
		(LACS)	= listener active state (L function)

*See the first paragraph of Section 2.12.3.7

the device to start a new message generation cycle.

In SWNS the SH function may send the DAV message true or false. In this state the device may change the multiline message being sent.

The SH function shall exit SWNS and enter:

(1) The SGNS if the nba message is false

(2) The SIWS within t_2 if either:

(a) The ATN message is true and neither CACS nor CTRS is active

(b) Or the ATN message is false and neither TACS nor SPAS is active.

2.3.3.6 *SIWS (Source Idle Wait State).* In SIWS the SH function is not active in the external message byte transfer process but is active in the internal process of waiting for the device to start a new message generation cycle. This SIWS allows a sequence of message byte transfers to be interrupted without loss of data over the interface while at the same time the device may continue to prepare for the new (next) message byte generation cycle.

In SIWS the DAV message shall be sent passive false.

The SH function shall exist SIWS and enter:

(1) The SIDS if the nba message is false

(2) The SWNS if either:

(a) The TACS is active

(b) Or the SPAS is active

(c) Or the CACS is active

2.3.4 SH Function Allowable Subsets. The only allowable subsets to the SH function shall be those listed in Table 4.

2.3.5 Additional SH Function Requirements and Guidelines. The nba true message indicates the device has generated a (new) message byte and made it available on the interface signal lines.

The nba message shall become true only in SIDS or SGNS. The nba message may become false in any other SH states.

The expression for interruption $(ATN \wedge \overline{(CACS) \vee (CTRS)}) \vee (\overline{ATN} \wedge \overline{(TACS) \vee (SPAS)})$ may be substituted by $\overline{(TACS)} \wedge \overline{(SPAS)} \wedge \overline{(CACS)} \wedge \overline{(CTRS)}$ if the transition of the latter expression can be effected within t_2 after the change of ATN.

2.4 AH (Acceptor Handshake) Interface Function

2.4.1 General Description. The AH function provides a device with the capability to guarantee proper reception of remote multiline messages. An interlocked handshake sequence between an SH function and one or more AH functions (each contained within separate devices) guarantees asynchronous transfer of each message byte. An AH function may delay either the initiation of, or termination of, a multiline message transfer until prepared to continue with the transfer process. The AH function utilizes the DAV, RFD, and DAC messages to effect each message byte transfer.

Table 6
AH Message Outputs

AH State	Remote Message Sent RFD	DAC	Device Function (DF) Interaction
AIDS	(T)	(T)	DF cannot receive remote multiline or END messages
ANRS	F	F	DF cannot receive remote multiline or END messages
ACRS	(T)	F	DF cannot receive remote multiline or END messages
AWNS	F	(T)	DF cannot receive remote multiline or END messages
ACDS	F	F	DF can receive remote multiline or END messages if LACS is active

2.4.2 AH Function State Diagram. The AH interface function shall be implemented so as to perform according to the state diagram given in Fig 4 and the state descriptions given throughout 2.4. Table 5 specifies the set of messages and states required to effect transition from one active state to another. Table 6 specifies the messages that shall be sent and the device function interaction required while each state is active.

2.4.3 AH Function State Descriptions

2.4.3.1 *AIDS (Acceptor Idle State).* In AIDS the AH function is inactive and not engaged in the handshake cycle. The AH function powers on in AIDS.

In AIDS the RFD and DAC messages shall be sent passive true.

The AH function shall exit AIDS and enter the ANRS (acceptor not ready state) within t_2 if either:

(1) The ATN message is true
(2) Or LACS is active
(3) Or LADS is active

2.4.3.2 *ANRS (Acceptor Not Ready State).* In ANRS the AH function indicates to the interface it has not yet prepared internally to continue with the handshake cycle.

In ANRS the RFD and DAC messages shall be sent false.

The AH Function shall exit ANRS and enter:

(1) The ACRS if the tcs (take control synchronously) message is false (see the first

Fig 4
AH State Diagram

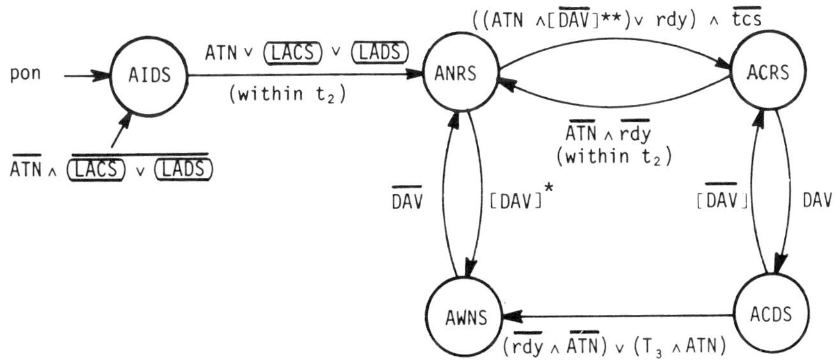

*This transition will never occur under normal interface operation; however, it may be implemented to simplify the interface function design.

**The need for the optional DAV term arises only for devices exiting AIDS and then only under take control synchronous (tcs) conditions (see 2.12.3.7). A minor revision to the controller function is under consideration that eliminates the need for the [DAV] term.

The addition of [$\overline{\text{DAV}}$] prevents a data byte from being misinterpreted as an interface message (due to momentary coincidence of DAV and ATN) when ANRS is entered from AIDS. If ANRS has been active for more than 1.5 μs, it can be assumed DAV is false.

paragraph of 2.12.3.7) and either:

(a) The ATN message is true and the DAV message is false

(b) Or the rdy (ready for next message) message is true

NOTE: Use of the DAV message is optional.

(2) The AIDS if the ATN message is false and neither:

(a) The LADS is active

(b) Nor LACS is active

(3) The AWNS if, optionally, the DAV message is true (note that this transition will never occur under normal interface operation)

2.4.3.3 *ACRS (Acceptor Ready State).* In ACRS the AH function indicates to the interface that it is prepared to receive multiline messages.

In ACRS the DAC message shall be sent false and the RFD message shall be sent passive true.

The AH function shall exit ACRS and enter:

(1) The ACDS (accept data state) if the DAV message is true

(2) The AIDS if the ATN message is false and neither:

(a) The LADS is active

(b) Nor LACS is active

(3) The ARNS within t_2 if both the ATN and the rdy messages are false

2.4.3.4 *ACDS (Accept Data State).* In ACDS the AH function indicates to the SH function that it shall maintain a valid message byte. This is the only state in which multiline messages on the DIO signal lines are valid. The ACDS indicates to the interface functions that an interface message is present and valid if the ATN message is true. The ACDS indicates to the device functions that a device dependent message is present and valid if LACS is active.

In ACDS the DAC and RFD messages shall be sent false.

The AH function shall exit the ACDS and enter:

(1) The AWNS (acceptor wait for new cycle state) if either:

(a) The ATN message is true and a period of T_3 has elapsed

(b) Or the ATN and rdy messages are both false

(2) The AIDS if the ATN message is false and neither:

(a) The LADS is active

(b) Nor LACS is active

(3) The ACRS if, optionally, the DAV message is false (note that this transition can occur only when the controller takes control asynchronously)

2.4.3.5 *AWNS (Acceptor Wait for New Cycle State).* In AWNS the AH function indicates that is has received a multiline message byte.

In AWNS the RFD message shall be sent false and the DAC message shall be sent passive true.

The AH function shall exit the AWNS and enter:

(1) The ANRS if DAV is false

(2) The AIDS if the ATN message is false and neither:

(a) The LADS is active

(b) Nor LACS is active

2.4.4 Acceptor Function Allowable Subsets. The only allowable subsets to the AH function shall be those listed in Table 7.

2.4.5 Additional AH Function Requirements and Guidelines. The local message rdy shall not become false during the ACRS. The transition from ACRS to ANRS shall occur only at the time ATN becomes false.

The RFD message received by an SH function is the logical AND of all the RFD messages sent by all the active AH functions. Similarly, the DAC message received by an SH function is the logical AND of all the DAC messages sent by all the AH functions. The way in which the composite effects of multiple AH functions interact with an SH function to perform the logical AND function via the use

Table 7
Allowable Subsets to AH Function

Identification	Description	States Omitted	Other Requirements	Other Function Subsets Required
AHØ	no capability	all	none	none
AH1	complete capability	none	none	none

of the NRFD and NDAC signal lines is explained further in 5.4.

Since interface functions need be designed only so as to perform according to the state diagrams specified, it is not required that exactly the states specified are the ones which exist in an implementation. One consequence of this statement is that interface function state transitions which are qualified by interface messages can occur after the message has been received as long as the RFD message is held false until they occur. The resulting performance cannot be distinguished from the performance of the specified diagrams in which the transitions shall occur while the interface message is being received. If this type of implementation is chosen, then the AH function should remain in ANRS even though the exit condition is true in order to hold the RFD message false (this is allowed by the last paragraph in 2.1.2.4).

In noisy environments, filter action on the incoming DAV message within a device can minimize false transitions to the ACDS state.

2.5 T (Talker) Interface Function (Includes Serial Poll Capabilities)

2.5.1 General Description. The T interface function provides a device with the capability to send device dependent data (including status data during a serial poll sequence) over the interface to other devices. This capability exists only when the T interface function is addressed to talk.

There are two alternative versions of the function: one with and one without address extension. The normal T function uses a 1 byte address. The T interface function with address extension [hereinafter called a TE (extended talker) function] uses a 2 byte address. In all other respects, the capabilities of both versions are the same.

Only one of the two alternative T functions need be implemented in a specific device.

NOTE: Both the T function and TE function are described concurrently throughout 2.5 due to the extensive similarity between these two functions.

2.5.2 T Function State Diagrams. The T function shall be implemented so as to perform according to the state diagrams given in Fig 5 and the state descriptions given throughout 2.5. Table 8 specifies the set of messages and states required to effect transition from one active state to another. Table 9 specifies the messages that shall be sent and the device function interaction required while each state is active.

The TE function shall be implemented so as to perform according to the state diagrams given in Fig 6 and the state descriptions given

Fig 5
T State Diagram

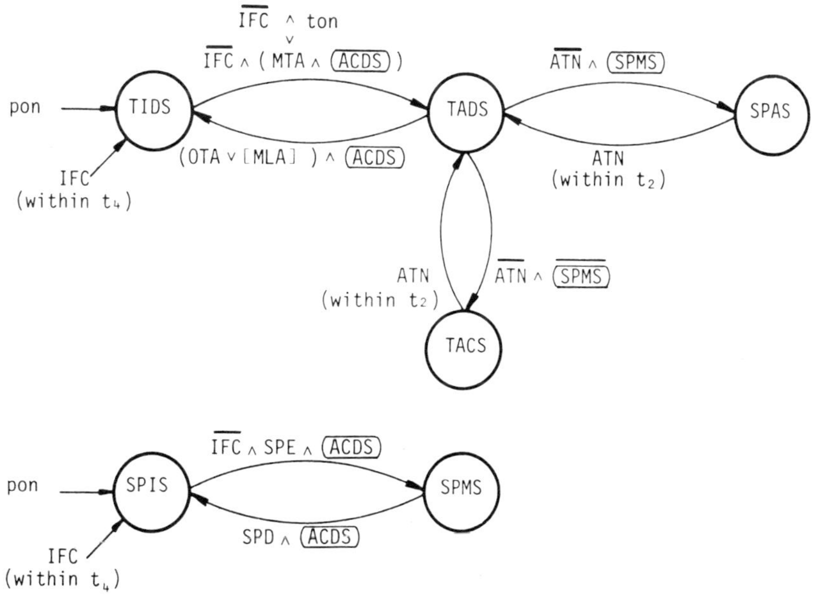

Table 8
T Mnemonics

Messages		Interface States	
pon	= power on	TIDS	= talker idle state
ton	= talk only	TADS	= talker addressed state
IFC	= interface clear	TACS	= talker active state
ATN	= attention	SPAS	= serial poll active state
MTA	= my talk address	SPIS	= serial poll idle state
SPE	= serial poll enable	SPMS	= serial poll mode state
SPD	= serial poll disable	(ACDS)	= accept data state (AH function)
OTA	= other talk address		
MLA	= my listen address		

Table 9
T or TE Message Outputs

T State	Qualifier	Remote Messages Sent* Multiline	END	RQS‖	Device Function (DF) Interaction
TIDS		(NUL)	(F)	(F)	DF not allowed to send messages
TADS		(NUL)	(F)	(F)	DF not allowed to send messages
TACS		DAB† or EOS†	T or F†	(F)	DF can send DAB, EOS, or END message (if used) concurrent with DAB‡
SPAS	APRS inactive	STB†	F or T	F	DF can send one STB message‡
SPAS	APRS active	STB†	F or T	T	DF can send one STB message‡

*See Table 38, Section 2.13.

†Messages enabled by the T function originating within the device functions.

‡ Under SH control.

‖See 2.5.3.4.

Table 10
TE Mnemonics

Messages		Interface States	
pon	= power on	TIDS	= talker idle state
ton	= talk only	TADS	= talker addressed state
IFC	= interface clear	TACS	= talker active state
ATN	= attention	SPAS	= serial poll active state
MTA	= my talk address	TPIS	= talker primary idle state
OTA	= other talk address	TPAS	= talker primary addressed state
OSA	= other secondary address	SPIS	= serial poll idle state
PCG	= primary command group	SPMS	= serial poll mode state
SPE	= serial poll enable	(ACDS)	= accept data state (AH function)
SPD	= serial poll disable	(LPAS)	= listener primary addressed state (L function)
MSA	= my secondary address		

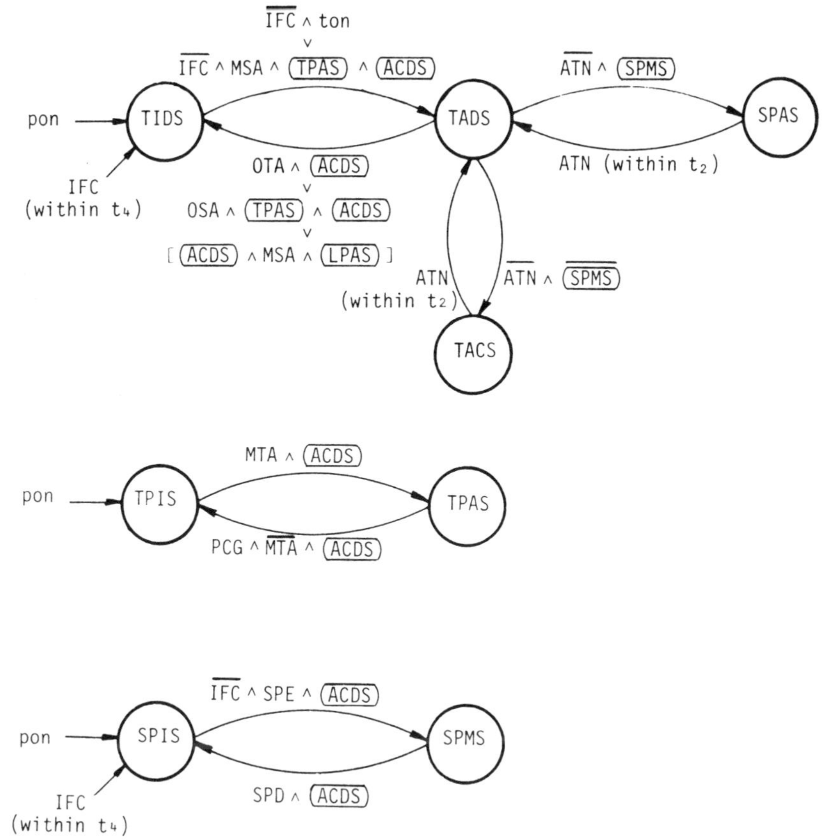

NOTE: If the TE function is used together with the L function instead of the LE function, then [MSA ∧ (ACDS) ∧ (LPAS)] shall be replaced by [MLA ∧ (ACDS)].

Fig 6
TE State Diagram

throughout 2.5. Table 10 specifies the set of messages and states required to effect transition from one active state to another. Table 9 specifies the messages that shall be sent and the device function interaction required while each state is active.

2.5.3 T Function State Descriptions

2.5.3.1 *TIDS (Talker Idle State).* In TIDS neither the T function nor the TE function is engaged in sending data or status bytes. The T function or the TE function powers on in TIDS.

In TIDS the END and RQS (request service) messages shall be sent passive false and the NUL message shall be sent passive true.

The T function shall exit the TIDS when the IFC messsage is false and enter the TADS (talker addressed state) if either:

(1) The MTA (my talk address) message is true and ACDS is active

(2) Or the ton (talk only) message is true (see the last paragraph of 2.5.5)

The TE function shall exit TIDS and enter TADS if the IFC message is false and either:

(1) The MSA (my secondary address) message is true, and ACDS is active, and the TPAS (talker primary address state) is active

(2) Or the ton message is true

2.5.3.2 *TADS (Talker Addressed State).* In TADS the T function has received its talk address and is prepared for, but not engaged in, sending data or status bytes. In TADS the TE function has received both its primary and secondary talk addresses and is prepared for, but not engaged in, sending data or status bytes.

In TADS the END and RQS messages shall be sent passive false and the NUL message shall be sent passive true.

The T function shall exit TADS and enter:

(1) The TACS (talker active state) if the ATN message is false and the SPMS (serial poll mode state) is inactive

(2) The SPAS (serial poll active state) if the ATN message is false and SPMS is active

(3) The TIDS if either:

(a) The OTA (other talk address) message is true and ACDS is active

(b) Or the MLA message is true and ACDS is active

(c) Or within t_4 if the IFC message is true

NOTE: Use of the MLA message is optional.

The TE function shall exit TADS and enter:

(1) The TACS is the ATN message is false and SPMS is inactive

(2) The SPAS is the ATN message is false and SPMS is active

(3) The TIDS if either:

(a) The OTA message is true and ACDS is active

(b) Or the OSA (other secondary address) message is true and TPAS and ACDS are active

(c) Or the MSA message is true and both the LPAS (listener primary addressed state) and ACDS are active

(d) Or within t_4 if the IFC message is true.

NOTE: Use of the expression containing the MSA message is optional.

2.5.3.3 *TACS (Talker Active State).* In TACS the T function, or the TE function, enables the transfer of the DAB (date byte) message and END, if used, from the device function to the interface signal lines. The message content is determined solely by the device function(s). The SH function determines when the device function(s) may change the message content of DAB (and END if used).

During TACS the DAB or EOS (end of string) and END messages may be sent by the device functions. The RQS message shall be sent passive false.

NOTE: The coding and format of the data is, in general, device dependent and beyond the scope of this standard.

The T function or the TE function shall exit TACS and enter:

(1) The TADS within t_2 if the ATN message is true

(2) The TIDS within t_4 if the IFC message is true

2.5.3.4 *SPAS (Serial Poll Active State).* In SPAS the T function or the TE function enables the transfer of a single status message from the device function to the interface signal lines using the SH function to control the transfer of the status byte that contains both the RQS and STB messages.

Although a controller needs only one byte for the STB and RQS messages from a device, it is allowable for the device to repeat this combined message byte if the controller does not assert ATN after the first transfer. In this case the content of the STB message may change between subsequent transfers although the RQS message is held unaltered by the SR function.

During SPAS, whether APRS state is active or inactive, the END message shall be sent either true or false. The RQS message shall be sent true if APRS is active, or false if APRS is inactive. In addition, the STB message shall be sent by the device function(s).

NOTE: The APRS is contained in the SR interface function.

The T function or the TE function shall exit SPAS and enter:

(1) The TADS within t_2 if the ATN message is true

(2) The TIDS within t_4 if the IFC message is true

2.5.3.5 *SPIS (Serial Poll Idle State).* In SPIS the T function or the TE function is not enabled to participate in a serial poll. The T or TE function powers on in SPIS.

The SPIS does not provide a remote message sending capability.

The T function or the TE function shall exit SPIS and enter SPMS if the SPE (serial poll enable) message is true and ACDS is active and the IFC message is false.

2.5.3.6 *SPMS (Serial Poll Mode State).* In SPMS the T function or the TE function is enabled to participate in a serial poll.

The SPMS does not provide a remote message sending capability.

The T function or the TE function shall exit SPMS and enter SPIS if either:

(1) The SPD (serial poll disable) message is true and the ACDS is active

(2) Or within t_4 if the IFC message is true

2.5.3.7 *TPIS (Talker Primary Idle State).* In TPIS the TE function is able to recognize its primary address and not able to respond to its secondary address. The TE function powers on in TPIS.

The TPIS does not provide a remote message sending capability.

The TE function shall exit TPIS and enter TPAS if the MTA message is true and ACDS is active.

2.5.3.8 *TPAS (Talker Primary Addressed State).* In TPAS the TE function is able to recognize and respond to its secondary address.

The TPAS does not provide a remote message sending capability.

The TE function shall exit TPAS and enter TPIS if the PCG (primary command group) message is true, the MTA message is false, and ACDS is active.

2.5.4 T Function and TE Function Allowable Subsets. The only allowable subsets to the T and TE interface functions shall be those listed in Tables 11 and 12.

2.5.5 Additional T and TE Interface Function Requirements and Guidelines. Each device which includes a T function or TE function shall provide a means by which the talk address (or secondary address) which it recognizes as MTA (or MSA) can be changed in the field by the user of the device.

The interruption of a device sending data by transitions in and out of TACS should not adversely affect the format of the output data. It is recommended that a device returning to TACS should continue with the output data string at the point of interruption.

Each device which includes the ton message shall be provided with a manual switch to generate the talk only condition.

2.6 L (Listener) Interface Function

2.6.1 General Description. The L interface function provides a device with the capability to receive device dependent data (including status data) over the interface from other devices. This capability exists only when the function is addressed to listen.

There are two alternative versions of the function: one with and one without address extension. The normal L function uses a 1 byte address. The L function with address extension [hereinafter called an LE (extended listener) function] uses a 2 byte address. In all other respects, the capabilities of both versions are the same.

Only one of the two alternative L functions need by implemented in a specific device.

Table 11
Allowable Subsets to T Interface Function

Identification	Description				States Omitted	Other Requirements	Other Function Subsets Required
	Capabilities						
	Basic Talker	Serial Poll	Talk Only Mode	Unaddress If MLA			
T0	N	N	N	N	all	none	none
T1	Y	Y	Y	N	none	omit [MLA $\wedge$ (ACDS)]	SH1 and AH1
T2	Y	Y	N	N	none	omit [MLA $\wedge$ (ACDS)] ton always false	SH1 and AH1
T3	Y	N	Y	N	SPIS, SPMS, SPAS	omit [MLA $\wedge$ (ACDS)]	SH1 and AH1
T4	Y	N	N	N	SPIS, SPMS, SPAS	omit [MLA $\wedge$ (ACDS)] ton always false	SH1 and AH1
T5	Y	Y	Y	Y	none	include [MLA $\wedge$ (ACDS)]	SH1 and L1-L4 or LE1-LE4
T6	Y	Y	N	Y	none	include [MLA $\wedge$ (ACDS)] ton always false	SH1 and L1-L4 or LE1-LE4
T7	Y	N	Y	Y	SPIS, SPMS, SPAS	include [MLA $\wedge$ (ACDS)]	SH1 and L1-L4 or LE1-LE4
T8	Y	N	N	Y	SPIS, SPMS, SPAS	include [MLA $\wedge$ (ACDS)] ton always false	SH1 and L1-L4 or LE1-LE4

Table 12
Allowable Subsets to TE Interface Function

Identification	Description: Capabilities — Basic Extended Talker	Serial Poll	Talk Only Mode	Unaddress If MSA ∧ (LPAS)	States Omitted	Other Requirements	Other Function Subsets Required
TE0	N	N	N	N	all	none	none
TE1	Y	Y	Y	N	none	omit [MSA ∧ (LPAS) ∧ (ACDS)]	SH1 and AH1
TE2	Y	Y	N	N	none	omit [MSA ∧ (LPAS) ∧ (ACDS)] ton always false	SH1 and AH1
TE3	Y	N	Y	N	SPIS, SPMS, SPAS	omit [MSA ∧ (LPAS) ∧ (ACDS)]	SH1 and AH1
TE4	Y	N	N	N	SPIS, SPMS, SPAS	omit [MSA ∧ (LPAS) ∧ (ACDS)] ton always false	SH1 and AH1
TE5	Y	Y	Y	Y	none	include [MSA ∧ (LPAS) ∧ (ACDS)]	SH1 and L1-L4 or LE1-LE4
TE6	Y	Y	N	Y	none	include [MSA ∧ (LPAS) ∧ (ACDS)] ton always false	SH1 and L1-L4 or LE1-LE4
TE7	Y	N	Y	Y	SPIS, SPMS, SPAS	include [MSA ∧ (LPAS) ∧ (ACDS)]	SH1 and L1-L4 or LE1-LE4
TE8	Y	N	N	Y	SPIS, SPMS, SPAS	include [MSA ∧ (LPAS) ∧ (ACDS)] ton always false	SH1 and L1-L4 or LE1-LE4

NOTE: Both the L function and LE function are described concurrently throughout 2.6 due to the extensive similarity between these two functions.

2.6.2 L Function State Diagram. The L function shall be implemented so as to perform according to the state diagram given in Fig 7 and the state descriptions given throughout 2.6. Table 13 specifies the set of messages and states required to effect transition from one active state to another. Table 14 describes the device function interaction required while each state is active.

The LE function shall be implemented so as to perform according to the state diagram in Fig 8 and the state descriptions given throughout 2.6. Table 15 specifies the set of messages and states required to effect transition from one active state to another. Table 14 describes the device function interaction required while each state is active.

2.6.3 L Function State Descriptions

2.6.3.1 *LIDS (Listener Idle State).* In LIDS neither the L function nor the LE function is engaged in the transfer of device dependent messages. The L or LE function powers on in the LIDS state.

The LIDS does not provide a remote message sending capability.

The L function shall exit LIDS and enter the LADS (listener addressed state) if the IFC message is false and either:

(1) The MLA (my listen address) message is true and ACDS is active

(2) Or the lon (listen only) message is true (see 2.6.5)

(3) Or the ltn (listen) message is true and CACS is active

The LE function shall exit LIDS and enter LADS if the IFC message is false and either:

(1) The MSA (my secondary address) message is true and the ACDS state is active, and the LPAS (listener primary addressed state) is active

(2) Or the lon message is true

(3) Or the ltn message is true and CACS is active

2.6.3.2 *LADS (Listener Addressed State).* In LADS the L function has received its listen address and is prepared for, but not engaged in, the transfer of device dependent messages. In LADS the LE function has received both its primary and secondary listen addresses and is prepared for, but not engaged in, the transfer of device dependent messages.

The LADS does not provide a remote message sending capability.

The L function shall exit LADS and enter:

(1) The LACS (listener active state) within t_2 if the ATN message is false

(2) The LIDS if either:

(a) The UNL (unlisten) message is true and ACDS is active

(b) Or the lun (local unlisten) message is true and CACS is active

(c) Or the MTA message is true and ACDS is active

(d) Or within t_4 if the IFC message is true

NOTE: Use of the MTA message is optional.

The LE function shall exit LADS and enter:

(1) The LACS within t_2 if the ATN message is false

Fig 7
L State Diagram

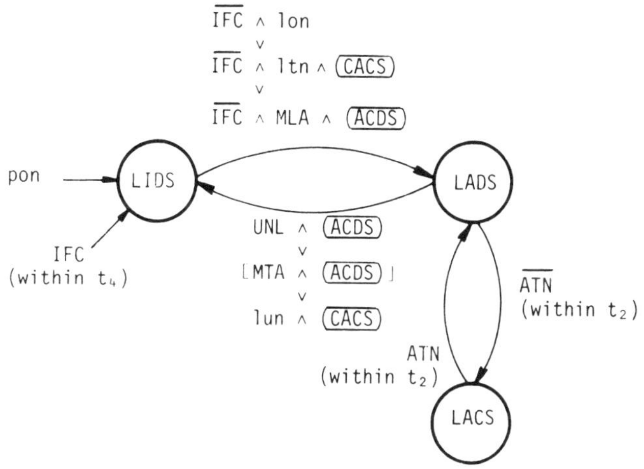

Table 13
L Mnemonics

Messages	Interface States
pon = power on	LIDS = listener idle state
ltn = listen	LADS = listener addressed state
lun = local unlisten	LACS = listener active state
lon = listen only	
IFC = interface clear	(ACDS) = accept data state (AH function)
ATN = attention	(CACS) = controller active state (C function)
UNL = unlisten	
MLA = my listen address	
MTA = my talk address	

Table 14
L or LE Message Outputs

L or LE State	Remote Messages Sent	Device Function (DF) Interaction
LIDS	none	DF not allowed to receive messages
LADS	none	DF not allowed to receive messages
LACS	none	DF can receive one device dependent message byte each time ACDS is active

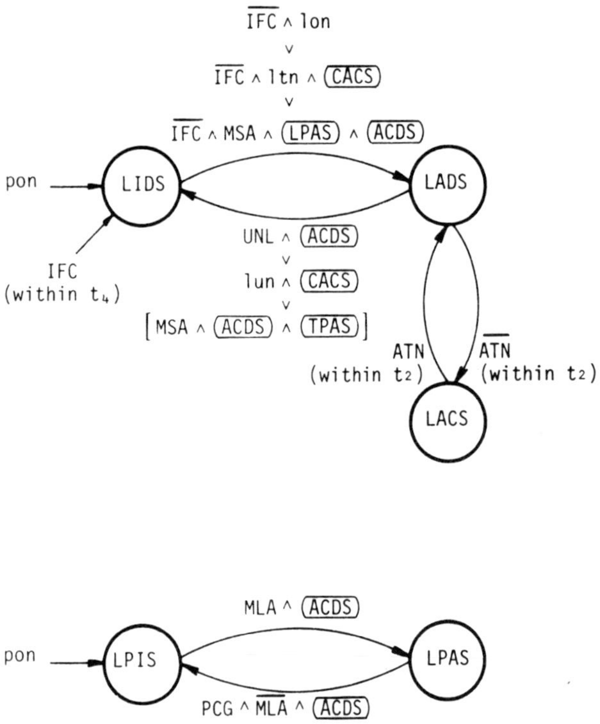

NOTE: If the LE function is used together with the T function, then [MSA ∧ (ACDS) ∧ (TPAS)] shall be replaced by [MTA ∧ (ACDS)].

Fig 8
LE State Diagram

(2) The LIDS if either:

(a) The UNL message is true and ACDS is active

(b) Or the lun message is true and CACS is active

(c) Or the MSA message is true and TPAS and ACDS are active

(d) Or within t_4 if the IFC message is true

NOTE: Use of the expression containing the MSA message is optional.

2.6.3.3 *LACS (Listener Active State).* In LACS the L function, or the LE function, is enabled to transfer any device dependent message (DAB, EOS, STB, END, or RQS) to the device functions as received via the interface signal lines. The AH function is used by the device function(s) to control the message transfer.

NOTE: The coding and format of the data is, in general, device dependent and beyond the scope of this standard.

The LACS does not provide a remote message sending capability.

The L function or the LE function shall exit LACS and enter:

(1) The LADS within t_2 if the ATN message is true

(2) The LIDS within t_4 if the IFC message is true

2.6.3.4 *LPIS (Listener Primary Idle State).* In LPIS the LE function is able to recognize its primary address and not able to respond to its secondary address. The LE function powers on in LPIS.

The LPIS does not provide a remote message sending capability.

The LE function shall exit LPIS and enter LPAS if the MLA message is true and ACDS is active.

2.6.3.5 *LPAS (Listener Primary Addressed State).* In LPAS the LE function is able to

Table 15
LE Mnemonics

Messages		Interface States	
pon	= power on	LIDS	= listener idle state
ltn	= listen	LACS	= listener active state
lun	= local unlisten	LADS	= listener addressed state
lon	= listen only	LPIS	= listener primary idle state
		LPAS	= listener primary addressed state
IFC	= interface clear		
ATN	= attention	(ACDS)	= accept data state (AH function)
UNL	= unlisten	(CACS)	= controller active state (C function)
MLA	= my listen address	(TPAS)	= talker primary addressed state (T function)
PCG	= primary command group		
MSA	= my secondary address		

recognize and respond to its secondary address.

The LPAS does not provide a remote message sending capability.

The LE function shall exit LPAS and enter LPIS if the PCG (primary command group) message is true, the MLA message is false, and ACDS is active.

2.6.4 L Function and LE Function Allowable Subsets. The only allowable subsets to the L and LE interface functions shall be those listed in Tables 16 and 17.

2.6.5 Additional L or LE Requirements and Guidelines. Each device which includes an L function or LE function shall provide a means by which the listen address and secondary address which it recognizes as MLA and MSA can be changed in the field by the user of the device.

The interruption of a device receiving data by transitions in and out of LACS should not adversely affect the future receipt of input data. It is recommended that a device re-

Table 16
Allowable Subsets to L Interface Function

Identification	Description: Capabilities: Basic Listener	Description: Capabilities: Listen Only Mode	Description: Capabilities: Unaddress If MTA	States Omitted	Other Requirements	Other Function Subsets Required
L∅	N	N	N	all	none	none
L1	Y	Y	N	none	omit [MTA ∧ (ACDS)]	AH1
L2	Y	N	N	none	omit [MTA ∧ (ACDS)] lon always false	AH1
L3	Y	Y	Y	none	include [MTA ∧ (ACDS)]	AH1 and T1-T8 or TE1-TE8
L4	Y	N	Y	none	include [MTA ∧ (ACDS)] lon always false	AH1 and T1-T8 or TE1-TE8

Table 17
Allowable Subsets to LE Interface Function

Identification	Description: Capabilities: Basic Extended Listener	Description: Capabilities: Listen Only Mode	Description: Capabilities: Unaddress If (MSA ∧ (TPAS))*	States Omitted	Other Requirements	Other Function Subsets Required
LE∅	N	N	N	all	none	none
LE1	Y	Y	N	none	omit [MSA ∧ (TPAS) ∧ (ACDS)]	AH1
LE2	Y	N	N	none	omit [MSA ∧ (TPAS) ∧ (ACDS)] lon always false	AH1
LE3	Y	Y	Y	none	include [MSA ∧ (TPAS) ∧ (ACDS)]	AH1 and T1-T8 or TE1-TE8
LE4	Y	N	Y	none	include [MSA ∧ (TPAS) ∧ (ACDS)] lon always false	AH1 and T1-T8 or TE1-TE8

*Replaced by MTA when used together with the T function.

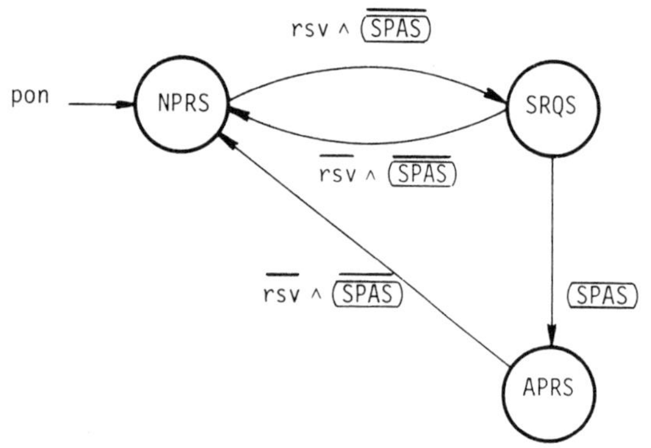

Fig 9
SR State Diagram

turning to LACS should continue with the input data string at the point of interruption.

Each device which includes the lon message shall be provided with a manual switch to generate the listen only condition.

2.7 SR (Service Request) Interface Function

2.7.1 General Description. The SR interface function provides a device with the capability to request service asynchronously from the controller in charge of the interface.

It also synchronizes the content of the RQS message of the composite status byte present during a serial poll so that the SRQ message can be removed from the interface once the RQS message is received true by the controller in charge (see 2.12.1).

2.7.2 SR Interface Function State Diagrams. The SR interface function shall be implemented so as to perform according to the state diagram given in Fig 9 and the state descriptions given throughout 2.7. Table 18 specifies the set of messages and states required to effect transition from one active state to another. Table 19 specifies the messages that shall be sent and the device function interaction required while each state is active.

2.7.3 SR State Description

2.7.3.1 *NPRS (Negative Poll Response State.)* In NPRS the SR function is not requesting service. The SR function powers on in NPRS.

Table 18
SR Mnemonics

Messages	Interface States
pon = power on	NPRS = negative poll response state
rsv = request service	SRQS = service request state
	APRS = affirmative poll response state
	(SPAS) = serial poll active state (T function)

Table 19
SR Message Outputs

SRQ State	Remote Message Sent: SRQ	Device Function Interaction
NPRS	(F)	none
SRQS	T	none
APRS	(F)	none

Table 20
Allowable Subsets to SR Interface Function

Identification	Description	States Omitted	Other Requirements	Other Function Subsets Required
SR∅	no capability	all	none	none
SR1	complete capability	none	none	T1, T2, T5, T6, TE1, TE2, TE5 or TE6

In NPRS the SRQ message shall be sent passive false.

NOTE: The RQS message will be sent false when SPAS is active (see 2.5.3.4) and NPRS is active.

The SR function shall exit NPRS and enter SRQS at any time the rsv (request service) message is true and SPAS is not active.

2.7.3.2 *SRQS (Service Request State).* In SRQS the SR function continuously indicates over the interface that it is requesting service.

In SRQS the SRQ message shall be sent true

The SR function shall exit SRQS and enter:

(1) The NPRS if the rsv message is false and SPAS is not active

(2) The APRS (affirmative poll response state) if SPAS is active

2.7.3.3 *APRS (Affirmative Poll Response State).* In APRS the SR function requires service but is not actively requesting it over the interface.

In APRS the SRQ message shall be sent passive false.

NOTE: The RQS message will be sent true by the talker when SPAS is active (see 2.5.3.4) and APRS is active.

The SR function shall exit APRS and enter NPRS at any time the rsv message is false and SPAS is not active.

2.7.4 SR Interface Function Allowable Subsets. The only allowable subsets to the SR interface function shall be those listed in Table 20.

2.7.5 Additional SR Interface Function Requirements and Guidelines. The SR function is required for each unique reason for requesting service.

If more than one reason exists, within a device, to request service, then a separate SR function and corresponding rsv message shall be used for each separate reason.

An SRQ true message should be sent when requested by any of the SR functions within a device.

While the T function is in the SPAS, the RQS message shall be sent true if any of the SR functions, within a device, is in the APRS. When the SR function exits SRQS, the SRQ message is not sent again until either the rsv message goes false and reoccurs, or a different SR function in the same device enters SRQS.

The SRQ message received, via the C (controller) function, is the logical OR of the SRQ messages sent by all SR functions. The way this is performed via the use of the SRQ signal line is explained in 5.4.2.

2.8 RL (Remote Local) Interface Function

2.8.1 General Description. The RL interface function provides a device with the capability to select between two sources of input information. The function indicates to the device that either input information from the front panel controls (local) or corresponding input information from the interface (remote) is to be used.

2.8.2 RL Function State Diagram. The RL interface function shall be implemented so as to perform according to the state diagram given in Fig 10 and the state descriptions given throughout 2.8. Table 21 specifies the set of

Fig 10
RL State Diagram

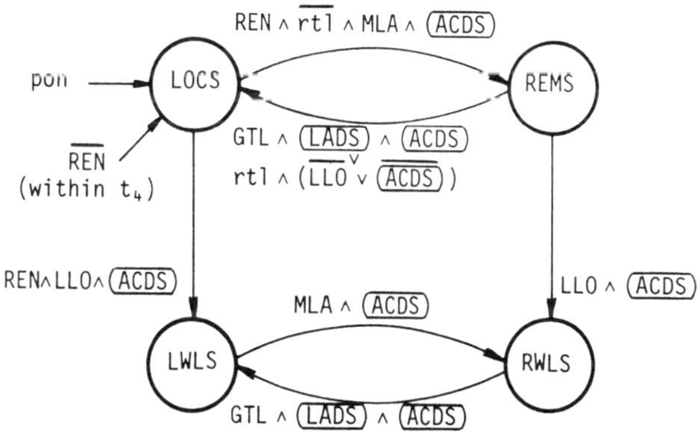

NOTE: If the RL function is used together with the LE function, then the term MLA shall be replaced by the term (MSA ∧ (LPAS)).

Table 21
RL Mnemonics

Messages		Interface States	
pon	= power on	LOCS	= local state
rtl	= return to local	LWLS	= local with lockout state
REN	= remote enable	REMS	= remote state
LLO	= local lockout	RWLS	= remote with lockout state
GTL	= go to local	(ACDS)	= accept data state (AH)
MLA	= my listen address	(LADS)	= listener addressed state (L)

messages and states required to effect transition from one active state to another. Table 22 specifies the device function interaction required while each state is active.

2.8.3 RL State Descriptions

2.8.3.1 *LOCS (Local State).* In LOCS all local controls of the associated device functions (front or rear panel) are operative and the device may store, but not respond to, corresponding device dependent messages from the interface. The RL function powers on in LOCS.

The LOCS does not provide a remote message sending capability.

The RL function shall exit LOCS if the REN message is true and enter:

(1) The REMS (remote state) if the rtl (return to local) message is false and the MLA message is true and ACDS is active

(2) The LWLS (local with lockout state) if the universal coded command LLO (local lockout) is true and ACDS is active

2.8.3.2 *LWLS (Local with Lockout State).* In LWLS all local device controls of the associated device functions are operative, and the device may store, but not respond to, corresponding device dependent messages from the interface (the rtl message is ignored).

The LWLS does not provide a remote message sending capability.

The RL function shall exit LWLS and enter:

(1) The RWLS (remote with lockout state) when MLA is true and ACDS is active

(2) The LOCS within t_4 if the REN message is false

2.8.3.2 *REMS (Remote State).* In the REMS state local controls (of the associated device functions) which have corresponding remote controls, except those controls which send local messages to interface functions, are inoperative and the device functions are under control of a remote device.

The REMS does not provide a remote message sending capability.

The RL function shall exit REMS and enter:

(1) The RWLS if the LLO message is true and ACDS is active

(2) The LOCS:

(a) Within t_4 if the REN message is false

(b) Or the GTL (go to local) message is true and ACDS and LADS are active

(c) Or the rtl message is true and either the LLO message is false or ACDS is inactive

2.8.3.4 *RWLS (Remote with Lockout State).* In RWLS those local controls (of the associated device functions) that have corresponding remote controls, except those controls that send local messages to interface functions, are inoperative and the device functions are under control of a remote device (the rtl message is ignored).

The RWLS does not provide a remote message sending capability.

The RL function shall exit RWLS and enter:

(1) The LOCS within t_4 if the REN message is false

(2) The LWLS if the GTL message is true and LADS and ACDS are active

2.8.4 RL Function Allowable Subsets. The only allowable subsets to the Remote Local

Table 22
RL Message Outputs

RL State	Remote Messages Sent	Device Function Interaction
LOCS	none	device is in "local control" mode
LWLS	none	device is in "local control" mode
REMS	none	device is in "remote control" mode
RWLS	none	device is in "remote control" mode

Table 23
Allowable Subsets to RL Interface Function

Identification	Description	States Omitted	Other Requirements	Other Function Subsets Required
RL∅	no capability	all	none	none
RL1	complete capability	none	none	L1-L4 or LE1-LE4
RL2	no local lock out	LWLS and RWLS	rtl always false	L1-L4 or LE1-LE4

Interface Function shall be those listed in Table 23.

2.8.5 Additional RL Interface Function Requirements and Guidelines. The ability of a device either to send device dependent messages over the interface or to receive and utilize device dependent messages not in conflict with locally available data is independent of the state which is active within the RL function.

When either REMS or RWLS is active, the associated device must ignore future use of local controls and become responsive to all subsequent input data received via the interface. However, as a function of its particular program storage capabilities, a device may be designed to either:

(1) Use its current local control settings until they are overridden by subsequent input data received via the interface

(2) Or use input data previously received via the interface

Conversely, when either LOCS or LWLS becomes active, the associated device shall become responsive to future use of local controls and ignore subsequent input data received via the interface. However, a device may be designed to either:

(1) Use the most recently received input data until it is overridden by subsequent local control settings

(2) Or use the current settings at the time when REMS or RWLS becomes inactive

It is required that the rtl message shall not be generated permanently.

Applications that require absolute local control of a device by a local programming source (for example, a human operator) are beyond the scope of this standard.

2.9 PP (Parallel Poll) Interface Function

2.9.1 General Description. The PP interface function provides a device with the capability to present a PPR message to the controller in charge without being previously addressed to talk.

The signal lines DIO1 through DIO8 are used to convey the device status bits during the parallel poll. In order for a device to respond with a PPR message, a device shall have been assigned (configured) to a single DIO line either by the controller or by a local message. This allows up to eight devices with a one-line-per-device assignment, although any number of devices can be handled through sharing of DIO lines.

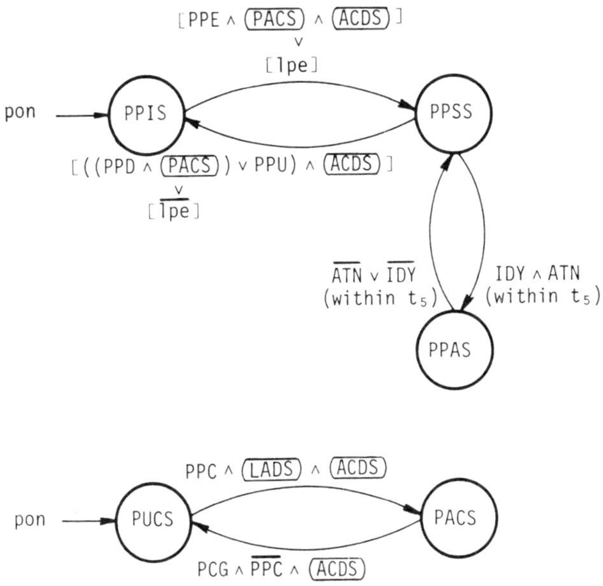

NOTE: See Table 27 for restrictions on use of the optional transitions.

Fig 11
PP State Diagram

The use of the parallel poll facility within a system requires a commitment of the current interface controller to conduct a parallel poll, as required.

The parallel poll facility can be used to indicate a request for service. This capability differs from use of the SRQ message in the following ways:

(1) A controller initiates a parallel poll sequence whereas any device requests the initiation of a serial poll sequence.

(2) A parallel poll enables the transfer of status data from multiple devices concurrently whereas a serial poll sequentially collects status data from each device.

2.9.2 PP Function State Diagram. The PP interface function shall be implemented so as to perform according to the state diagram given in Fig 11 and the state descriptions given throughout 2.9. Table 24 specifies the set of messages and states required to effect transition from one active state to another. Table 25 specifies the messages that shall be sent and the device function interaction required by the function while each state is active.

2.9.3 PP State Descriptions

2.9.3.1 *PPIS (Parallel Poll Idle State).* In PPIS the PP function is unable to respond to a parallel poll issued by the interface controller. The PP function powers on in PPIS.

In PPIS all PPR (parallel poll response) messages shall be sent passive false.

The PP function shall exit PPIS and enter the PPSS (parallel poll standby state) if either:

(1) The PPE (parallel poll enable) message is true and PACS and ACDS are active

(2) Or the lpe (local poll enabled) message is true

NOTE: Both the lpe and PPE transitions are optional; only one shall be used at any given time.

2.9.3.2 *PPSS (Parallel Poll Standby State).* In PPSS the PP function is able to respond to parallel polls issued by the device controller whenever they occur.

In PPSS all PPR messages shall be sent passive false.

The PP function shall exit the PPSS and enter:

(1) The PPAS (parallel poll active state) within t_5 if the IDY (identify) and ATN messages are true (a parallel poll is in progress)

(2) The PPIS if:

(a) The lpe message is false

(b) Or the PPD (parallel poll disable) message is true and PACS and ACDS are active

(c) Or the PPU (parallel poll unconfigure) message is true and ACDS is active

NOTE: Both the lpe and PPD transitions are optional; only one shall be used at any given time.

2.9.3.3 *PPAS (Parallel Poll Active State).* In PPAS the PP function is responding to the parallel poll currently being conducted by the interface controller.

In PPAS one of the PPR messages shall be sent true if, and only if, the value of the ist (individual status) message is equal to the value of the S (sense) bit received as part of the most recently received PPE command. The PPR message to be sent shall be the one specified by the three bits P1 through P3 received as part of the most recently received PPE command. Table 26 lists the PPR message specified by each of the combinations of values of

Table 24
PP Mnemonics

Messages		Interface States	
pon	= power on	PPIS	= parallel poll idle state
ist	= individual status (Table 25)	PPSS	= parallel poll standby state
lpe	= local poll enabled	PPAS	= parallel poll active state
ATN	= attention	PUCS	= parallel poll unaddressed to configure state
IDY	= identify	PACS	= parallel poll addressed to configure state
PPE	= parallel poll enable	(ACDS)	= accept data state (AH function)
PPD	= parallel poll disable	(LADS)	= listener addressed state (L function)
PPC	= parallel poll configure		
PCG	= primary command group		
PPU	= parallel poll unconfigure		

Table 25
PP Message Outputs

PP State	Qualifier	Remote Message Sent PPRn*†	Device Function Interfaction
PPIS		(F)	none
PPSS		(F)	none
PPAS	ist ≡ S†	T	none
PPAS	ist ≢ S†	(F)	none

*This column refers only to the specific message assigned by the device.

†See 2.9.3.3, second paragraph.

Table 26
PPR Message Specified by Each of the Combinations of Values P1 Through P3

Bits Received with Most Recent PPE Command			PPR Message Specified
P3	P2	P1	
0	0	0	PPR1
0	0	1	PPR2
0	1	0	PPR3
0	1	1	PPR4
1	0	0	PPR5
1	0	1	PPR6
1	1	0	PPR7
1	1	1	PPR8

P1 through P3 (see 2.9.5). All other PPR messages should be sent passive false.

The PP interface function shall exit PPAS and enter PPSS within t_5 if either the IDY or ATN message is false (the parallel poll is over).

2.9.3.4 *PUCS (Parallel Poll Unaddressed to Configure State).* In PUCS the PP function shall ignore any PPE or PPD messages which might be received over the interface. The PP function powers on in PUCS.

The PUCS does not provide a remote message sending capability.

The PP function shall exit PUCS and enter the PACS (parallel poll addressed to configure state) if the PPC message is true, and LADS and ACDS are active.

2.9.3.5 *PACS (Parallel Poll Addressed to Configure State).* In PACS the PP function is able to act upon PPE or PPD messages received over the interface. If a PPE message is received, the attendant bits S, P1, P2, and P3 should be saved by the function.

The PACS does not provide a remote message sending capability.

The PP function shall exit PACS and enter PUCS when the PCG message is true, the PPC (parallel poll configure) message is false, and ACDS is active.

2.9.4 PP Interface Function Allowable Subsets. The only allowable subsets to the parallel poll interface function shall be those listed in Table 27.

2.9.5 Additional PP Interface Function Requirements and Guidelines. If subset PP2 is taken, field-settable local messages shall substitute for the PPE command to specify the PPR message and the message sense to be used during a parallel poll.

2.10 DC (Device Clear) Interface Function

2.10.1 General Description. The DC interface function provides the device with the capability to be cleared (initialized) either individually or as part of a group of devices. The group may be either a subset or all addressed devices in one system.

2.10.2 DC Function State Diagram. The DC interface function shall be implemented so as to perform according to the state diagram given in Fig 12 and the state descrip-

Fig 12
DC State Diagram

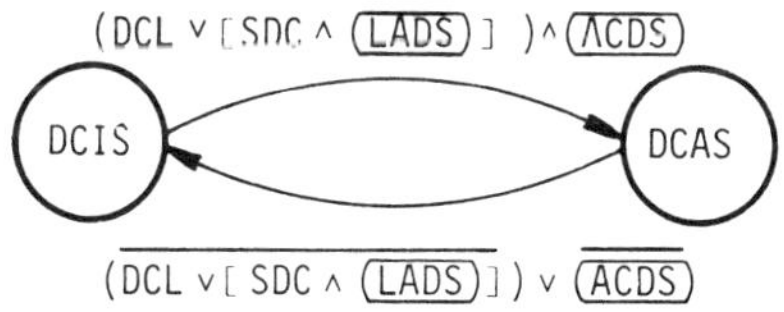

Table 27
Allowable Subsets to PP Interface Function

Identification	Description	States Omitted	Other Requirements	Other Function Subsets Required
PP∅	no capability	all	none	none
PP1	remote configuration	none	include [((PPD ∧ (PACS)) ∨ PPU) ∧ (ACDS)] include [PPE ∧ (PACS) ∧ (ACDS)] exclude lpe	L1-L4 or LE1-LE4
PP2	local configuration	PUCS, PACS	include lpe exclude [((PPD ∧ (PACS)) ∨ PPU) ∧ (ACDS)] exclude [PPE ∧ (PACS) ∧ (ACDS)] local messages shall be substituted for S, P1, P2, P3	none

tions given throughout 2.10. Table 28 specifies the set of messages and states required to effect transition from one active state to another. Table 29 specifies the device function interaction required while each state is active.

2.10.3 DC Function State Descriptions

2.10.3.1 *DCIS (Device Clear Idle State).* In DCIS the DC function is inactive.

The DCIS does not provide a remote message sending capability.

The DC function shall exit DCIS and enter the DCAS (device clear active state) if ACDS is active, and either:

(1) The DCL (device clear) message is true

(2) Or the SDC (selected device clear) message is true and LADS is active

NOTE: Use of the expression containing the SDC message is optional.

2.10.3.2 *DCAS (Device Clear Active State).* In DCAS the DC function sends an internal message to the device function(s) causing it (them) to be cleared.

The DCAS does not provide a remote message sending capability.

The DC function shall exit DCAS and enter the DCIS (device clear idle state) if either ACDS is inactive or neither:

(1) The DCL message is true

(2) Nor the SDC message is true and LADS is active

NOTE: Use of the expression containing the SDC message is optional.

2.10.4 DC Interface Function Allowable Subsets. The only allowable subsets to the DC interface function shall be those listed in Table 30.

2.10.5 Additional DC Function Requirements and Guidelines. The DCAS affects only device functions and does not affect other interface functions (cleared by IFC).

A device may use the DC function for any purpose consistent with its operation. Normal use of this function would place the device functions in the pon (power-on state). However, this function may be used to put any subset of the devices' functions to a defined state deemed appropriate by the designer, which state the designer shall then specify.

Table 28
DC Mnemonics

Messages	Interface States
DCL = device clear	DCIS = device clear idle state
SDC = selected device clear	DCAS = device clear active state
	(ACDS) = accept data state (AH function)
	(LADS) = listener addressed state (L function)

Table 29
DC Message Outputs

DC State	Remote Message Sent	Device Function (DF) Interaction
DCIS	none	normal device function operation
DCAS	none	DF should return to a known fixed state

Table 30
Allowable Subsets to DC Interface Function

Identification	Description	States Omitted	Other Requirements	Other Function Subsets Required
DCØ	no capability	all	none	none
DC1	complete capability	none	none	L1-L4 or LE1-LE4
DC2	omit selective device clear	none	omit [SDC ∧ (LADS)]	AH1

2.11 DT (Device Trigger) Interface Function.

2.11.1 General Description. The DT interface function provides the device with the capability to have its basic operation started either individually or as part of a group of devices. The group may be either a subset or all addressed devices in one system.

2.11.2 DT Function State Diagram. The DT interface function shall be implemented so as to perform according to the state diagram given in Fig 13 and the state descriptions given throughout 2.11. Table 31 specifies the set of messages and states required to effect transition from one active state to another. Table 32 specifies the device function interaction required while each state is active.

2.11.3 DT Function State Descriptions

2.11.3.1 *DTIS (Device Trigger Idle State).* In DTIS the DT function is inactive. The DTIS does not provide a remote message sending capability.

The DT function shall exit DTIS and enter the DTAS (device trigger active state) if:

(1) The GET (group execute trigger) message is true

(2) And LADS and ACDS are active

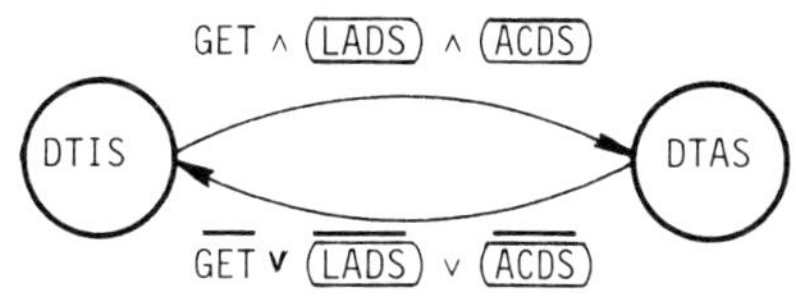

Fig 13
DT State Diagram

2.11.3.2 *DTAS (Device Trigger Active State).* In DTAS the DT function sends an internal message to the device function causing it to start performing its basic operation.

The DTAS does not provide a remote message sending capability.

The DT function shall exit DTAS and enter DTIS if either:

(1) The GET message is false

(2) Or LADS is inactive

(3) Or ACDS is inactive

2.11.4 DT Interface Allowable Subsets. The only allowable subsets to the DT interface function shall be those listed in Table 33.

Table 31
DT Mnemonics

Messages	Interface States	
GET = group execute trigger	DTIS	= device trigger idle state
	DTAS	= device trigger active state
	(ACDS)	= accept data state (AH function)
	(LADS)	= listener addressed state (L function)

Table 32
DT Message Outputs

DT State	Remote Messages Sent	Device Function (DF) Interaction
DTIS	none	normal DF operation
DTAS	none	DF should start performing triggered operation

Table 33
Allowable Subsets to DT Interface Function

Identification	Description	States Omitted	Other Requirements	Other Function Subsets Required
DTØ	no capability	all	none	none
DT1	complete capability	none	none	L1-L4 or LE1-LE4

2.11.5 Additional DT Function Requirements and Guidelines. The DTAS indicates that the device (or defined portions of the device) is to start performing its designated operation.

It is recommended that the device should begin the operation immediately after DTAS becomes active.

Once a device operation has been started, it shall not respond to subsequent state transitions until the operation is complete. Only after completion of the first operation can the device start a new operation in response to the next DTAS active condition.

2.12 C (Controller) Interface Function

2.12.1 General Description. The C interface function provides a device with the capability to send device addresses, universal commands and addressed commands to other devices over the interface. It also provides the capability to conduct parallel polls to determine which devices require service.

A C interface function can exercise its capabilities only when it is sending the ATN message over the interface.

If more than one device on the interface has a C interface function, then all but one of them shall be in the CIDS (controller idle state) at any given time. The device containing the C interface function which is not in the CIDS is called the controller-in-charge (of the interface system). Protocol is provided within this standard to allow devices with a C interface function to take turns as the controller-in-charge of the interface.

The C interface function in one of the devices connected to an interface (but no more than one) can exist in the SACS (system control active state). It shall remain in this state throughout operation of the interface and so possesses the capability to send the IFC and REN messages over the interface at any time whether or not it is the controller-in-charge. This device is called the system controller (of the interface system).

2.12.2 C Function State Diagram. The C interface function shall be implemented so as to perform according to the state diagram given in Fig 14 and the state descriptions given throughout 2.12.1. Table 34 specifies the set of messages and states required to effect transition from one active state to another. Table 35 specifies the messages that shall be sent and the device function interaction required while each state is active.

Table 34
C Mnemonics

Messages		Interface States	
pon	= power on	CIDS	= controller idle state
rsc	= request system control	CADS	= controller addressed state
rpp	= request parallel poll	CTRS	= controller transfer state
gts	= go to standby	CACS	= controller active state
tca	= take control asynchronously	CPWS	= controller parallel poll wait state
tcs	= take control synchronously	CPPS	= controller parallel poll state
sic	= send interface clear	CSBS	= controller standby state
sre	= send remote enable	CSHS	= controller standby hold state
IFC	= interface clear	CAWS	= controller active wait state
ATN	= attention	CSWS	= controller synchronous wait state
TCT	= take control	CSRS	= controller service requested state
		CSNS	= controller service not requested state
		SNAS	= system control not active state
		SACS	= system control active state
		SRIS	= system control remote enable idle state
		SRNS	= system control remote enable not active state
		SRAS	= system control remote enable active state
		SIIS	= system control interface clear idle state
		SINS	= system control interface clear not active state
		SIAS	= system control interface clear active state
		(ACDS)	= accept data state (AH function)
		(ANRS)	= acceptor not ready state (AH function)
		(SDYS)	= source delay state (SH function)
		(STRS)	= source transfer state (SH function)
		(TADS)	= talker addressed state (T function)

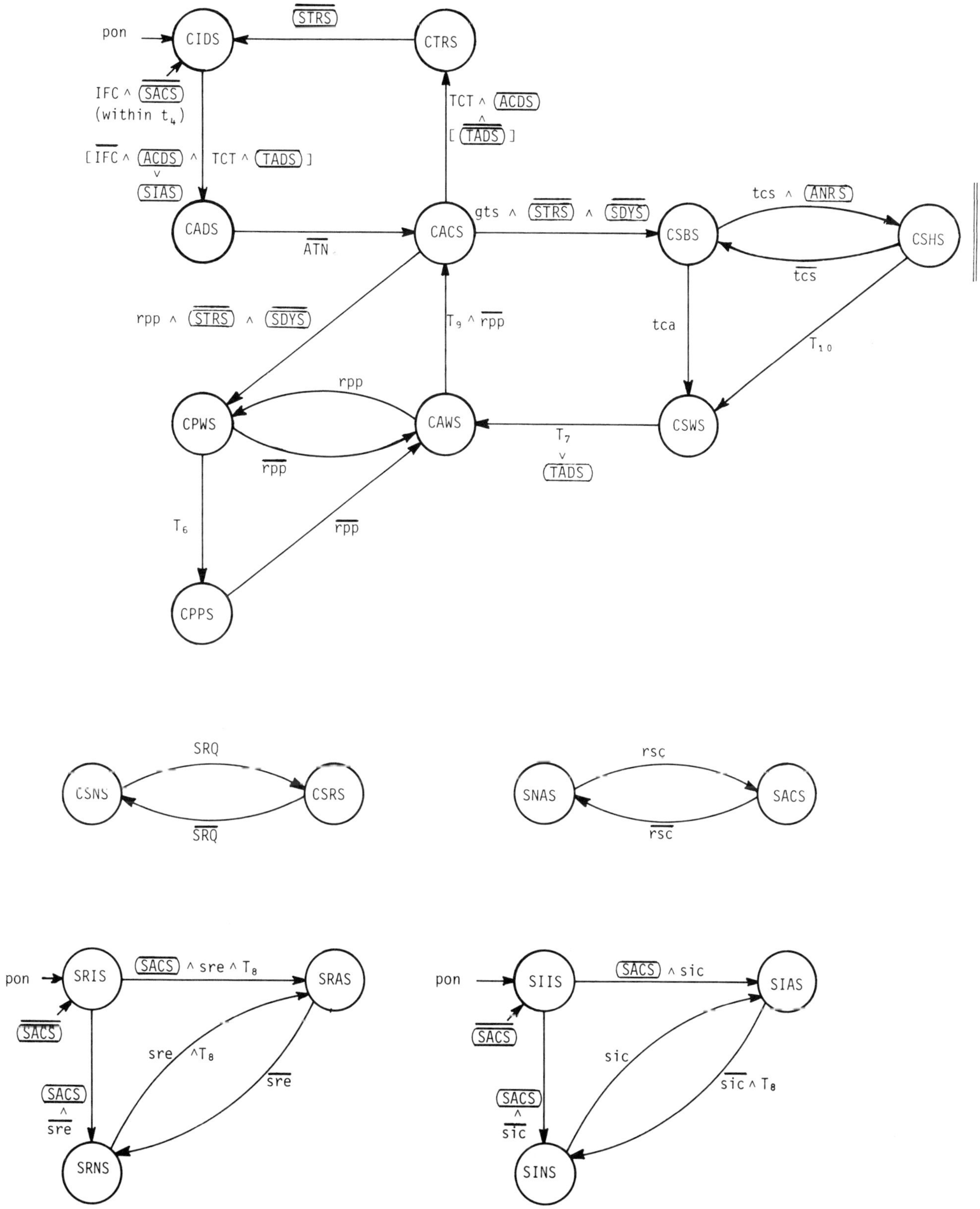

Fig 14
C State Diagram

2.12.3 C State Descriptions

2.12.3.1 *CIDS (Controller Idle State).* In CIDS the C function relinguishes all of its interface control capabilities. The C function powers on in CIDS.

In CIDS the ATN and IDY messages shall be sent passive false and the NUL message shall be sent passive true.

The C function shall exit CIDS and enter CADS (controller addressed state) when either:

(1) The TCT (take control) message (sent by the controller-in-charge) is true, the TADS and ACDS are active, and the IFC message is false

(2) Or the SIAS (system control interface clear active state) is active

NOTE: The expression containing the TCT message is optional.

2.12.3.2 *CADS (Controller Addressed State).* In CADS the C function is in the process of becoming the controller-in-charge of the interface but is waiting until the current controller stops sending the ATN message.

In CADS the ATN and IDY messages shall be sent passive false and the NUL message shall be sent passive true.

The C function shall exit CAS and enter:

(1) The CACS (controller active state) if the ATN message is false

(2) The CIDS within t_4 if the IFC message

Table 35
C Message Outputs*

C State	Remote Messages Sent: ATN	IDY	Multiline	Device Function (DF) Interaction
CIDS	(F)	(F)	(NUL)	DF shall not send interface messages
CADS	(F)	(F)	(NUL)	DF shall not send interface messages
CACS	T	F	†	DF can send interface messages
CPWS	T	T	(NUL)	DF shall not send interface messages
CPPS	T	T	(NUL)	DF can receive PPR messages
CSBS	F	(F)	(NUL)	DF shall not send interface messages
CSHS	F	(F)	(NUL)	DF shall not send interface messages
CSWS	T	F or (F)	(NUL)	DF shall not send interface messages
CAWS	T	F	(NUL)	DF shall not send interface messages
CTRS	T	F	TCT	DF shall finish sending TCT message

C State	Remote Messages Sent: IFC	Device Function (DF) Interaction
SIIS	(F)	none
SINS	F	none
SIAS	T	none

C State	Remote Messages Sent: REN	Device Function (DF) Interaction
SRIS	(F)	none
SRNS	F	none
SRAS	T	none

C State	Remote Messages Sent	Device Function (DF) Interaction
CSNS	none	no service requests exist
CSRS	none	DF notified of request for service

*Message values sent are shown opposite only those states which affect them. Each major section of the table corresponds to a group of mutually exclusive states within the controller function.

†Any coded interface message listed in Table 38. Although enabled by the C function, these messages originate within the device functions.

Table 36
Multiline Messages

Universal Commands (multiline)	Addresses
LLO	(LAD)*
DCL	(TAD)†
SPE	UNL
SPD	
PPU	
Addressed Commands	**Secondary Commands**
GET	(SAD)‡
GTL	PPD
PPC	PPE
SDC	
TCT	

*Represents a listen address of a specific device (received as MLA).

†Represents a talk address of a specific device (received as MTA or OTA).

‡Represents a secondary address of a specific device (received as MSA or OSA).

is true and SACS is not active

2.12.3.3 *CACS (Controller Active State).* In CACS the C function enables the transfer of multiline interface messages from the device function(s) to the interface signal lines. These messages include device addresses, universal commands, or addressed commands. The SH function determines when the device function(s) may change the message content of the multiline messages being sent. However, message content is determined solely by the device function(s).

The ATN message shall be sent continuously true and the IDY message shall be sent continuously false, while CACS is active, under which conditions any of the multiline messages in Table 36 may be sent by the device functions.

The C function shall exit CACS and enter:

(1) The CTRS (controller transfer state) if the TCT message is true, TADS is (optionally) inactive, and ACDS is active

(2) The CPWS (controller parallel poll wait state) if the rpp (request parallel poll) message is true, and neither SDYS nor STRS is active

(3) The CIDS within t_4 if the IFC message is true and SACS is not active

(4) The CSBS (controller standby state) if the gts (go to standby) message is true, and neither STRS nor SDYS is active.

2.12.3.4 *CPWS (Controller Parallel Poll Wait State).* In CPWS the C function is conducting a parallel poll over the interface but waiting for the DIO lines to settle.

In CPWS the ATN and IDY messages shall be sent true and the NUL message shall be sent passive true.

The C function shall exit CPWS and enter:

(1) The CPPS (controller parallel poll state) after a period of T_6 has elapsed

(2) The CIDS within t_4 if the IFC message is true and SACS is not active

(3) The CAWS state if the rpp message is false

2.12.3.5 *CPPS (Controller Parallel Poll State).* In CPPS the C function is conducting a parallel poll and actively transferring PPR message values to the device function(s) as received via the interface signal lines.

In CPPS the ATN and IDY messages shall be sent true and the NUL message shall be sent passive true.

The C function shall exit CPPS and enter:

(1) The CAWS if the rpp message is false

(2) The CIDS within t_4 if the IFC message is true and SACS is not active

2.12.3.6 *CSBS (Controller Standby State).* In CSBS the C function is allowing two or more devices to transfer device dependent messages over the interface.

In CSBS the ATN message shall be sent false, the IDY message shall be sent passive false, and the NUL message shall be sent passive true.

The C function shall exit CSBS and enter:

(1) The CSHS (controller standby hold state) if the tcs (take control synchronously) message is true and ANRS is active

(2) The CSWS (controller synchronous wait state) if the tca (take control asynchronously) message is true

(3) The CIDS within t_4 if the IFC message is true and SACS is not active

2.12.3.7 *CSWS (Controller Synchronous Wait State).* In CSWS the C function is in the process of entering the CAWS (controller active wait state) but is waiting for a specified time T_7 to make sure that the current active talker recognizes the ATN message being sent over the interface. If this state were entered via the tcs message, the device function(s) shall continue to send it true during this state.

This causes the AH interface function to continue sending the RFD message false over the interface, holding off transfer of the next data byte.

In CSWS the ATN message shall be sent true, the IDY message shall be sent active or passive false, and the NUL message shall be sent passive true.

The C function shall exit CSWS and enter:

(1) The CAWS after a period of T_7 has elapsed or if TADS is active

(2) The CIDS within t_4 if the IFC message is true and SACS is not active

2.12.3.8 *CAWS (Controller Active Wait State).* In CAWS the C function is waiting for a period of T_9 before entering CACS. This wait shall occur in order to guarantee that the EOI line has settled to its proper value and that no device is responding erroneously to what appears to be a parallel poll.

In CAWS the ATN message shall be sent true, the IDY message shall be sent false, and the NUL message shall be sent passive true.

The C function shall exit CAWS and enter:

(1) The CACS if the rpp message is false and a period of T_9 has elapsed

(2) The CPWS if the rpp message is true

(3) The CIDS within t_4 if the IFC message is true and SACS is not active

2.12.3.9 *CTRS (Controller Transfer State).* In CTRS the C function is sending the TCT addressed command to another device and is thus in the process of becoming idle.

In CTRS the ATN message shall be sent true, the IDY message shall be sent false, and the TCT message shall continue to be sent true.

The C function shall exit CTRS and enter CIDS when either:

(1) The STRS becomes inactive

(2) Or within t_4 if the IFC message is true and SACS is not active

2.12.3.10 *CSRS (Controller Service Requested State).* In CSRS the C function is notifying the device function(s) via a local message that at least one device on the interface is requesting service.

The CSRS does not provide a remote message sending capability.

The C function shall exit CSRS and enter the CSNS (controller service not requested state) if the SRQ message is false.

2.12.3.11 *CSNS (Controller Service Not Requested State).* In CSNS the C function is notifying the device function(s) via a local message that no device on the interface is requesting service.

The CSNS does not provide a remote message sending capability.

The C function shall exit CSNS and enter CSRS if the SRQ message is true.

2.12.3.12 *SNAS (System Control Not Active State).* In SNAS the C function relinguishes all of its system control capabilities. The SNAS does not provide a remote message sending capability.

The C function shall exit SNAS and enter SACS if the rsc (request system control) message is true.

2.12.3.13 *SACS (System Control Active State).* In SACS the C function is allowed to exercise its system control capabilities. The SACS does not provide a remote message sending capability.

The C function shall exit SACS and enter SNAS if the rsc message is false.

2.12.3.14 *SIIS (System Control Interface Clear Idle State).* In SIIS the C function has no capability to clear the interface. The C interface function powers on in SIIS.

In SIIS the IFC message shall be sent passive false.

The C function shall exit SIIS if SACS is active and enter:

(1) The SINS (system control interface clear not active state) if the sic (send interface clear) message is false

(2) The SIAS (system control interface clear active state) if the sic message is true

2.12.3.15 *SINS (System Control Interface Clear Not Active State).* In SINS the C function is not engaged in clearing the interface.

In SINS the IFC message shall continuously be sent false.

The C function shall exit SINS and enter:

(1) The SIAS if the local sic message is true

(2) The SIIS if SACS is not active

2.12.3.16 *SIAS (System Control Interface Clear Active State).* In SIAS the C function is engaged in clearing the interface.

All interface functions connected to the system shall respond to the IFC true message and will transfer to a known initial state.

In SIAS the IFC message shall be sent true.

The C function shall exit SIAS and enter:

(1) The SINS if the sic message is false and SIAS has been active for at least a period of T_8

(2) The SIIS if SACS is not active

2.12.3.17 *SRIS (System Control Remote Enable Idle State).* In SRIS the C function has no remote enable capability. All implementations of the C function should remain in the SRIS continuously except when used in a device capable of system controller performance. The C function powers on in the SRIS state.

In SRIS the REN messsage shall be sent passive false.

The C function shall exit SRIS and enter:

(1) The SRNS (system control remote enable not active state) if the sre (send remote enable) message is false and SACS is active

(2) The SRAS (system control remote enable state) if the sre message is true, SACS is active, and SRIS has been active for at least a period of T_8

2.12.3.18 *SRNS (System Control Remote Enable Not Active State).* In SRNS the C function is not engaged in enabling remote operation of other devices over the interface.

In SRNS the REN message shall be sent passive false.

The C function shall exit SRNS and enter:

(1) The SRAS if the sre message is true for at least a period of T_8

(2) The SRIS if SACS is not active

2.12.3.19 *SRAS (System Control Remote Enable Active State).* In SRAS the C function is actively engaged in enabling remote operations of other devices over the interface.

In SRAS the REN message shall be continuously sent true.

The C function shall exit SRAS and enter:

(1) The SRNS if the sre message is false

(2) The SRIS if SACS is not active

2.12.3.20 CSHS (Controller Standby Hold State). In CSHS the C function is in a hold state until such time as the DAV message is false as presented to all devices connected to the system. The CSHS state prevents false coincidence of the ATN and DAV messages, as observed by idle devices, during a tcs sequence.

In CSHS the ATN message shall be sent false, the IDY message shall be sent passive false, and the NUL message shall be sent passive true.

The C function shall exit CSHS and enter:

(1) The CSWS (Controller Standby Wait State) if a period of T_{10} has elapsed

(2) The CSBS if the tcs message is false

(3) The CIDS within t_4 if the IFC message is true and SACS is not active

2.12.4 C Interface Function Allowable Subsets. The only allowable subsets to the C interface function shall be those listed in Table 37.

2.12.5 Additional C Function Requirements and Guidelines. Warning: Use tca with caution.

Restriction on the use of tca: The designer shall not assume that valid data will be transferred across the interface if the tca message becomes true while a device dependent message is true.

Background: Asynchronous interruption of an active talker by a controller through the use of tca may occur at any time when a device dependent message is true. If a device dependent message is true and ATN becomes true, the interrupted byte could be misinterpreted by other devices as an interface message (for example, command or address) and produce unintended state transitions.

The tcs message, if used, may change from false to true only during CSBS. It may change from true to false only during the CAWS. These restrictions guarantee that RFD is held false for the proper amount of time during a synchronous take-control operation.

A device with system controller capability should be provided with a manual means to interrupt the local rsc message if that device is to be used in multiple controller environments.

2.13 Remote Message Coding and Transfer

2.13.1 Remote Message Coding. Each remote message is sent by an interface function or received by an interface function via one or more interface signal lines. This section defines the complete set of remote messages and how they are coded and transferred on the signal lines. The coding of all remote messages sent or received by the various interface functions is specified in Table 38, Remote Message Coding.

2.13.2 Remote Message Coding Concepts. Messages may be coded into the logical state of one or more signal lines.

For this standard, a message derived from or sent as the logical state of only one signal line is referred to as a uniline message (for example, ATN).

For this standard, a message derived from or sent as a combination of logical states of two or more signal lines is referred to as a multiline

Table 37
Allowable Subsets to Controller Interface Function

Identi-fication*	Capabilities										Notes	States Required									Other Requirements			Other Function Subsets Required				
	System Controller	Send IFC and Take Charge	Send REN	Respond to SRQ	Send I.F. Messages	Receive Control	Pass Control	Pass Control to Self	Parallel Poll	Take Control Synchronously		SNAS, SACS	SIIS, SIAS, SINS	SRIS, SRAS, SRNS	CSNS, CSRS	CACS, CSBS, CSHS, CSWS, CAWS	CADS	CIDS	CTRS	CPWS, CPPS	[TCT ∧ (ACDS) ∧ (TADS)]†	[(TADS)]‡	*tcs* not always false	C1	C2	AH1, L1-L4, or LE1-LE4	SH1	T1-T8, TE1-TE8
C0	N	N	N	N	N	N	N	N	N	N		O	O	O	O	O	O	O	O	O	O	O	O	O	—	—	—	—
C1	Y	—	—	—	—	—	—	—	—	—	(1)	R	—	—	—	—	—	—	—	—	—	—	—	—	—	—	—	—
C2	—	Y	—	—	—	—	—	—	—	—	(1),(6)	—	R	—	—	—	—	—	—	—	—	—	—	R	—	—	—	—
C3	—	—	Y	—	—	—	—	—	—	—	(1)	—	—	R	—	—	—	—	—	—	—	—	—	R	—	—	—	—
C4	—	—	—	Y	—	—	—	—	—	—	(1)	—	—	—	R	—	—	—	—	—	—	—	—	—	—	—	—	—
C5	—	—	—	—	Y	Y	Y	Y	Y	Y	(2),(3)	—	—	—	—	R	R	R	R	R	R	R	R	—	—	R	R	R
C6	—	—	—	—	Y	Y	Y	Y	Y	N	(2),(3)	—	—	—	—	R	R	R	R	R	R	R	O	—	—	—	R	R
C7	—	—	—	—	Y	Y	Y	Y	N	Y	(2),(3)	—	—	—	—	R	R	R	R	O	R	R	R	—	—	R	R	R
C8	—	—	—	—	Y	Y	Y	Y	N	N	(2),(3)	—	—	—	—	R	R	R	R	O	R	R	O	—	—	—	R	R
C9	—	—	—	—	Y	Y	Y	N	Y	Y	(2),(3)	—	—	—	—	R	R	R	R	R	R	O	R	—	—	R	R	R
C10	—	—	—	—	Y	Y	Y	N	Y	N	(2),(3)	—	—	—	—	R	R	R	R	R	R	O	O	—	—	—	R	R
C11	—	—	—	—	Y	Y	Y	N	N	Y	(2),(3)	—	—	—	—	R	R	R	R	O	R	O	R	—	—	R	R	R
C12	—	—	—	—	Y	Y	Y	N	N	N	(2),(3)	—	—	—	—	R	R	R	R	O	R	O	O	—	—	—	R	R
C13	—	—	—	—	Y	Y	N	N	Y	Y	(2)	—	—	—	—	R	R	R	O	R	O	O	R	—	R	R	R	—
C14	—	—	—	—	Y	Y	N	N	Y	N	(2)	—	—	—	—	R	R	R	O	R	O	O	O	—	R	—	R	—
C15	—	—	—	—	Y	Y	N	N	N	Y	(2)	—	—	—	—	R	R	R	O	O	O	O	R	—	R	R	R	—
C16	—	—	—	—	Y	Y	N	N	N	N	(2)	—	—	—	—	R	R	R	O	O	O	O	O	—	R	—	R	—
C17	—	—	—	—	Y	N	Y	Y	Y	Y	(2),(3),(4)	—	—	—	—	R	O	R	R	R	R	R	R	—	—	R	R	R
C18	—	—	—	—	Y	N	Y	Y	Y	N	(2),(3),(4)	—	—	—	—	R	O	R	R	R	R	R	O	—	—	—	R	R
C19	—	—	—	—	Y	N	Y	Y	N	Y	(2),(3),(4)	—	—	—	—	R	O	R	R	O	R	R	R	—	—	R	R	R
C20	—	—	—	—	Y	N	Y	Y	N	N	(2),(3),(4)	—	—	—	—	R	O	R	R	O	R	R	O	—	—	—	R	R
C21	—	—	—	—	Y	N	Y	N	Y	Y	(2),(3),(4)	—	—	—	—	R	O	R	R	R	R	O	R	—	—	R	R	R
C22	—	—	—	—	Y	N	Y	N	Y	N	(2),(3),(4)	—	—	—	—	R	O	R	R	R	R	O	O	—	—	—	R	R
C23	—	—	—	—	Y	N	Y	N	N	Y	(2),(3),(4)	—	—	—	—	R	O	R	R	O	R	O	R	—	—	R	R	R
C24	—	—	—	—	Y	N	Y	N	N	N	(2),(3),(4)	—	—	—	—	R	O	R	R	O	R	O	O	—	—	—	R	R
C25	—	—	—	—	Y	N	N	N	Y	Y	(2),(5)	—	—	—	—	R	O	O	O	R	O	O	R	—	—	R	R	—
C26	—	—	—	—	Y	N	N	N	Y	N	(2),(5)	—	—	—	—	R	O	O	O	R	O	O	O	—	—	—	R	—
C27	—	—	—	—	Y	N	N	N	N	Y	(2),(5)	—	—	—	—	R	O	O	O	O	O	O	R	—	—	R	R	—
C28	—	—	—	—	Y	N	N	N	N	N	(2),(5)	—	—	—	—	R	O	O	O	O	O	O	O	—	—	—	R	—

*Typical notation to describe a controller consists of the letter C followed by one or more of the numbers indicating the subsets selected. For example; C1, 2, 3, 4, 8.

†This is part of the CIDS to CADS transitional expression.

‡This is part of the CACS to CTRS transitional expression.

NOTES:
(1) One or more of subsets C1 through C4 may be chosen in any combination with any one of C5 through C28.
(2) Only one subset may be chosen from C5 through C28.
(3) The CTRS state shall be included in devices which are to be operated in multicontroller systems.
(4) These subsets are not allowed unless C2 is included.
(5) These subsets are intended to be used in devices and systems where no control passage is possible.
(6) When a system controller asserts IFC during the time another physical device is operating as controller-in-charge, the system controller should refrain from active assertion of the source handshake and ATN until the removal of the IFC message to preclude multiple controller contention.
O = omit, R = required, hyphen = not applicable or not required, Y = yes, N = no.

message (for example, DCL).

A message may be defined as a logical combination (AND, OR, or NOT) of other messages (for example, OTA).

The coding of a message sent and received is the same.

2.13.3 Remote Message Transfer. A message is sent by driving one or more specified signal lines to a logical 1 or a logical Ø. Lines not specified as part of the message coding shall not be driven.

A message is received by sensing one or more specified bus signal lines to determine the logical value of each signal line as either 1 or Ø. Lines not specified as part of the message coding are ignored.

A uniline message value is considered valid as soon as its corresponding logic state is detected. (See Tables 3, 6, 9, 19, 25, 35 for times at which messages may be sent.)

A multiline message is valid only within the context of the SH and AH functions. A transmitted multiline message is valid while the SH function is in the STRS (source transfer state). A received multiline message is valid while the AH function is in the ACDS (accept data state).

All passive message values are transferred as Ø signal line states. This requires only the logic OR of signal line states to be performed on the interface.

2.13.4 Remote Message Coding Table Organization and Conventions. All messages capable of being sent or received by an interface function are listed by name and mnemonic in Table 38.

The table correlates the message value (true or false) to the bus signal line logical value (1 or Ø) and vice versa.

Each remote message entry in the table specifies both the encoding required to send the messages and the decoding required to receive the messages.

The true value of a uniline message is specified by the assignment of a specific logical state to a signal line.

The true value of a multiline message is specified by the assignment of a unique set of logical states (1 or Ø) to the corresponding set of signal lines that contain the message.

The false value of a message is any combination of logic states (1 or Ø) other than the unique set that specifies the true value.

Each message entry in the table is identified by type: either uniline U or multiline M. Each message is further identified by class (1 or 7) according to the function it performs within the interface function or device function.

The logical state a bus signal line may have is specified in the table as a Ø, 1, Y, or X. These represent the logic states as follows:

Ø = logical zero
1 = logical one
X = don't care (for the coding of a received message)
X = shall not drive unless directed by another message (for the coding of a transmitted message)
Y = don't care (for the coding of a transmitted message)

2.13.5 Remote Message Coding Table Perspective. Table 38 is constructed to reflect each remote message sent (or received) by each interface function. In actual use, two or more messages as defined in the table may be sent concurrently (for example, DAB true and ATN false) by different interface functions. See Notes 9 and 10 to Table 38 and Appendix D.

2.13.6 Summary Notes and Symbols for Remote Message Coding Table 38.

Level Assignment:

Ø = High state signal level
1 = Low state signal level

The coding of Table 38 may be translated to equivalent electrical signal levels as specified in 3.2.

Symbols:

Type	U	= Uniline message
	M	= Multiline message
Class	AC	= Addressed command
	AD	= Address (talk or listen)
	DD	= Device dependent
	HS	= Handshake
	UC	= Universal command
	SE	= Secondary
	ST	= Status

2.13.7 ISO Code Representation: Message Coding Guidelines. Many devices use the ISO-7 bit code (or the equivalent code in American National Standard Code for Information Interchange, ANSI X3.4-1977) because it is convenient to both generate and interpret this code. The relationships between the ISO code and the messages (binary bit patterns) defined and described in this standard are identified in this clause.

Table 38
Remote Message Coding

Mnemonic	Message Name		Type	Class	DIO 8	7	6	5	4	3	2	DIO 1	DAV	NRFD	NDAC	ATN	EOI	SRQ	IFC	REN
					Bus Signal Line(s) and Coding That Asserts the True Value of the Message															
ACG	addressed command group		M	AC	Y	Ø	Ø	Ø	X	X	X	X	X	X	X	1	X	X	X	X
ATN	attention		U	UC	X	X	X	X	X	X	X	X	X	X	X	1	X	X	X	X
DAB	data byte	(Notes 1, 9)	M	DD	D8	D7	D6	D5	D4	D3	D2	D1	X	X	X	Ø	X	X	X	X
DAC	data accepted		U	HS	X	X	X	X	X	X	X	X	X	X	Ø	X	X	X	X	X
DAV	data valid		U	HS	X	X	X	X	X	X	X	X	1	X	X	X	X	X	X	X
DCL	device clear		M	UC	Y	Ø	Ø	1	Ø	1	Ø	Ø	X	X	X	1	X	X	X	X
END	end	(Note 9)	U	ST	X	X	X	X	X	X	X	X	X	X	X	Ø	1	X	X	X
EOS	end of string	(Notes 2, 9)	M	DD	E8	E7	E6	E5	E4	E3	E2	E1	X	X	X	Ø	X	X	X	X
GET	group execute trigger		M	AC	Y	Ø	Ø	Ø	1	Ø	Ø	Ø	X	X	X	1	X	X	X	X
GTL	go to local		M	AC	Y	Ø	Ø	Ø	Ø	Ø	Ø	1	X	X	X	1	X	X	X	X
IDY	identify		U	UC	X	X	X	X	X	X	X	X	X	X	X	X	1	X	X	X
IFC	interface clear		U	UC	X	X	X	X	X	X	X	X	X	X	X	X	X	X	1	X
LAG	listen address group		M	AD	Y	Ø	1	X	X	X	X	X	X	X	X	1	X	X	X	X
LLO	local lock out		M	UC	Y	Ø	Ø	1	Ø	Ø	Ø	1	X	X	X	1	X	X	X	X
MLA	my listen address	(Note 3)	M	AD	Y	Ø	1	L5	L4	L3	L2	L1	X	X	X	1	X	X	X	X
MTA	my talk address	(Note 4)	M	AD	Y	1	Ø	T5	T4	T3	T2	T1	X	X	X	1	X	X	X	X
MSA	my secondary address	(Note 5)	M	SE	Y	1	1	S5	S4	S3	S2	S1	X	X	X	1	X	X	X	X
NUL	null byte		M	DD	Ø	Ø	Ø	Ø	Ø	Ø	Ø	Ø	X	X	X	X	X	X	X	X
OSA	other secondary address		M	SE	(OSA = SCG ∧ $\overline{\text{MSA}}$)															
OTA	other talk address		M	AD	(OTA = TAG ∧ $\overline{\text{MTA}}$)															
PCG	primary command group		M	—	(PCG = ACG ∨ UCG ∨ LAG ∨ TAG)															
PPC	parallel poll configure		M	AC	Y	Ø	Ø	Ø	Ø	1	Ø	1	X	X	X	1	X	X	X	X
PPE	parallel poll enable	(Note 6)	M	SE	Y	1	1	Ø	S	P3	P2	P1	X	X	X	1	X	X	X	X
PPD	parallel poll disable	(Note 7)	M	SE	Y	1	1	1	D4	D3	D2	D1	X	X	X	1	X	X	X	X
PPR1	parallel poll response 1	(Note 10)	U	ST	X	X	X	X	X	X	X	1	X	X	X	1	1	X	X	X
PPR2	parallel poll response 2	(Note 10)	U	ST	X	X	X	X	X	X	1	X	X	X	X	1	1	X	X	X
PPR3	parallel poll response 3	(Note 10)	U	ST	X	X	X	X	X	1	X	X	X	X	X	1	1	X	X	X
PPR4	parallel poll response 4	(Note 10)	U	ST	X	X	X	X	1	X	X	X	X	X	X	1	1	X	X	X
PPR5	parallel poll response 5	(Note 10)	U	ST	X	X	X	1	X	X	X	X	X	X	X	1	1	X	X	X

(Continued on page 51)

Table 38 (Continued)

Bus Signal Line(s) and Coding That Asserts the True Value of the Message

Mnemonic	Message Name		Type	Class	DIO 8	7	6	5	4	3	2	DIO 1	DAV	NRFD	NDAC	ATN	EOI	SRQ	IFC	REN
PPR6	parallel poll response 6	(Note 10)	U	ST	X	X	1	X	X	X	X	X	X	X	X	1	1	X	X	X
PPR7	parallel poll response 7	(Note 10)	U	ST	X	1	X	X	X	X	X	X	X	X	X	1	1	X	X	X
PPR8	parallel poll response 8	(Note 10)	U	ST	1	X	X	X	X	X	X	X	X	X	X	1	1	X	X	X
PPU	parallel poll unconfigure		M	UC	Y	∅	∅	1	∅	1	∅	1	X	X	X	1	X	X	X	X
REN	remote enable		U	UC	X	X	X	X	X	X	X	X	X	X	X	X	X	X	X	1
RFD	ready for data		U	HS	X	X	X	X	X	X	X	X	X	∅	X	X	X	X	X	X
RQS	request service	(Note 9)	U	ST	X	1	X	X	X	X	X	X	X	X	X	∅	X	X	X	X
SCG	secondary command group		M	SE	Y	1	1	X	X	X	X	X	X	X	X	1	X	X	X	X
SDC	selected device clear		M	AC	Y	∅	∅	∅	∅	1	∅	∅	X	X	X	1	X	X	X	X
SPD	serial poll disable		M	UC	Y	∅	∅	1	1	∅	∅	1	X	X	X	1	X	X	X	X
SPE	serial poll enable		M	UC	Y	∅	∅	1	1	∅	∅	∅	X	X	X	1	X	X	X	X
SRQ	service request		U	ST	X	X	X	X	X	X	X	X	X	X	X	X	X	1	X	X
STB	status byte	(Notes 8, 9)	M	ST	S8	X	S6	S5	S4	S3	S2	S1	X	X	X	∅	X	X	X	X
TCT	take control		M	AC	Y	∅	∅	∅	1	∅	∅	1	X	X	X	1	X	X	X	X
TAG	talk address group		M	AD	Y	1	∅	X	X	X	X	X	X	X	X	1	X	X	X	X
UCG	universal command group		M	UC	Y	∅	∅	1	X	X	X	X	X	X	X	1	X	X	X	X
UNL	unlisten		M	AD	Y	∅	1	1	1	1	1	1	X	X	X	1	X	X	X	X
UNT	untalk	(Note 11)	M	AD	Y	1	∅	1	1	1	1	1	X	X	X	1	X	X	X	X

The 1/∅ coding on ATN when sent concurrent with multiline messages has been added to this revision for interpretive convenience.

NOTES:

(1) D1-D8 specify the device dependent data bits.

(2) E1-E8 specify the device dependent code used to indicate the EOS message.

(3) L1-L5 specify the device dependent bits of the device's listen address.

(4) T1-T5 specify the device dependent bits of the device's talk address.

(5) S1-S5 specify the device dependent bits of the device's secondary address.

(6) S specifies the sense of the PPR.

S	Response
∅	∅
1	1

P1-P3 specify the PPR message to be sent when a parallel poll is executed.

P3	P2	P1	PPR Message
∅	∅	∅	PPR1
.	.	.	.
.	.	.	.
.	.	.	.
1	1	1	PPR8

(7) D1-D4 specify don't-care bits that shall not be decoded by the receiving device. It is recommended that all zeroes be sent.

(8) S1-S6, S8 specify the device dependent status. (DIO7 is used for the RQS message.)

(9) The source of the message on the ATN line is always the C function, whereas the messages on the DIO and EOI lines are enabled by the T function.

(10) The source of the messages on the ATN and EOI lines is always the C function, whereas the source of the messages on the DIO lines is always the PP function.

(11) This code is provided for system use, see 6.3.

2.13.7.1 *Interface Messages.* The interface system utilizes message coding as defined in Table 38 to carry interface messages among devices when the ATN message is true. This coding may be correlated to the ISO code by relating DIO1 through DIO7 to bits 1 through 7, respectively. The ISO code does not contain the equivalent of the dedicated ATN message (bit or line).

When the interface system defined in this standard is interconnected, via a terminal unit, to other environments, then protocol beyond the scope of this standard shall be used to enable proper communication and avoid possible contradictions with other assigned meanings for the ISO code.

2.13.7.2 *Device Dependent Messages.* The specific coding of device dependent messages is beyond the scope of this standard. After a talker and listener(s) have been addressed via interface messages, any commonly understood binary, BCD, or alphanumeric code may be used when the ATN message is false.

(1) The alphanumeric codes (dense subset of the ISO code, columns 2 through 5) are preferred for communication of the device dependent messages wherever possible. Bit 1 through bit 7 of the ISO code corresponds to DIO1-DIO7.

(2) When other codes are used (for example, binary) the most significant bit should be placed on the DIO line that has the highest number (for example, DIO8 for bit 8).

The ISO code is further illustrated in Appendix E as it correlates with the codes of this standard.

2.13.8 State Transition Timing Values. The T_x and t_y values listed in Section 2 throughout the interface function descriptions and state diagrams are defined in 3.8.

3. Electrical Specifications

3.1 Application. This section defines the electrical specifications for interface systems to be used in environments where:

(1) Physical distance between devices is short

(2) Electrical noise is relatively low

All electrical specifications for the driver and receiver circuits are based on the use of TTL (transitor transitor logic) technology.

NOTES:

(1) Interface function circuitry connected to the drivers or receivers may be implemented in other device technologies at the designer's choice.

(2) Driver and receiver devices need only be used on those signal lines required for the interface functions implemented (see 3.5.1 for termination requirements).

(3) Either open collector or three-state drivers may be used as determined by data rate considerations of 3.3 and 5.2.

3.2 Logical and Electrical State Relationships. The relationship between the logical states defined in Table 38, Remote Message Coding, and the electrical state levels present on the signal lines is as follows:

Coding Logical State	Electrical Signal Levels
0	corresponds to $\geq +2.0$ V, called high state
1	corresponds to $\leq +0.8$ V, called low state

The high and low states are based on standard TTL levels for which the power source does not exceed +5.25 V dc and is referenced to logic ground.

This section indicates current flow into a node with a positive sign and current flow out of a node with a negative sign.

3.3 Driver Requirements. Messages may be sent in either an active or passive manner over the interface (see 2.1.3). All passive true message transfer occurs in the high state and shall be carried on a signal line using open collector drivers.

3.3.1 Driver Types. Open collector drivers shall be used to drive the SRQ, NRFD, and NDAC signal lines.

Open collector drivers or three-state drivers may be used to drive DIO 1-8, DAV, IFC, ATN, REN, and EOI signal lines with this exception: DIO1-8 shall use open collector drivers for parallel polling applications (see 2.9.3.3).

NOTE: Three-state drivers are useful for systems where higher speed operation is required.

It is recommended that a three-state driver be used within a controller to drive the ATN signal line if the controller is intended to be used in a system in which other devices are implemented with three-state drivers on the DIO, DAV, and EOI signal lines.

3.3.2 Driver Specifications. The specifications for drivers shall be as follows:

Low state: Output voltage (three-state or open collector drivers) $< +0.5$ V at +48 mA sink current

The driver shall be capable of sinking 48 mA continuously.

High state: Output voltage (three-state) $\geq +2.4$ V at -5.2 mA
Output voltage (open collector) (see 3.5)

The indicated voltage values are measured at the device connector between the signal line and logic ground.

See 3.5.3 for additional requirements which may apply to the driver.

3.4 Receiver Requirements

3.4.1 Receiver Specifications, Allowed. The specification for receivers with nominal noise immunity shall be as follows:

Low state: Input voltage $\leq +0.8$ V
High state: Input voltage $\geq +2.0$ V

See 3.5.3 for additional requirements which may apply to the receiver.

3.4.2 Receiver Specifications, Preferred. To provide added noise immunity, the use of Schmitt-type receiver circuits (or equivalent) for all signal lines is recommended. The specifications for these receivers shall be as follows:

Hysteresis: $V_t\text{pos} - V_t\text{neg} \geq +0.4$ V
Low state: Negative going threshold voltage $V_t\text{neg} \geq +0.8$ V
High state: Positive going threshold voltage $+2.0 \geq V_t\text{pos}$

3.5 Composite Device Load Requirements

3.5.1 Resistive Termination. Each signal line (whether or not it is connected to a driver or receiver) shall be terminated within the device by a resistive load whose major purpose is to establish a steady-state voltage when all drivers on a line are in the high-impedance state. This load is also used to maintain a uniform device impedance on the line and improve noise immunity. For specific requirements see the last paragraph of 3.5.3, and for typical resistive values see 3.5.5.

3.5.2 Negative Voltage Clamping. Each signal line to which a receiver is connected shall be provided with means to limit the negative voltage excursions. Typically this circuit element is a diode clamp contained within the receiver component.

3.5.3 DC Load Requirements. The dc load characteristics of a device are affected by the driver and receiver circuits as well as the resistive termination and voltage clamping circuits; therefore they are specified for the composite device interface circuit, not for the individual components. This section, however, provides complete specifications for the resistive termination and voltage clamping circuits.

Load measurement conditions assume that the receiver, driver, and resistive termination circuits are connected together within the device with the driver in the high-impedance state.

Each signal line interface within a device shall have the following dc load characteristics and shall fall within the unshaded area of Fig 15.

(1) if $I \leq 0$ mA, V shall be < 3.7 V
(2) if $I \geq 0$ mA, V shall be > 2.5 V
(3) if $I \geq -12.0$ mA, V shall be > -1.5 V (only if receiver exists)
(4) if $V \leq 0.4$ V, I shall be < -1.3 mA
(5) if $V \geq 0.4$ V, I shall be > -3.2 mA
(6) if $V \leq 5.5$ V, I shall be $<$ 2.5 mA
(7) if $V \geq 5.0$ V, I shall be $>$ 0.7 mA or the small-signal Z shall be $\leq$ 2 kΩ at 1 MHz

3.5.4 Capacitive Load Limit. The internal capacitance load on each signal line shall not exceed 100 pF within each device (see 5.2).

NOTE: The effect of device capacitance on bus operations is most critical at low voltages. Since the design of driver and receiver circuits may contribute capacitive loads that vary with voltage, the capacitance should be measured at several voltage levels, all below 2 V, with the device powered on.

3.5.5 Typical Circuit Configuration. Fig 16 shows a typical circuit configuration for signal line input-output circuits for which readily available components exist.

This basic circuit is compatible with both TTL microcircuit and discrete element devices.

The specifications for this typical configuration are as follows:

R_{L1}: 3 kΩ ±5 percent (to V_{cc})
R_{L2}: 6.2 kΩ ±5 percent (to ground)
Driver: Output leakage current (open collector driver)
+0.25 mA max at V_0 = +5.25 V
Output leakage current (three-state driver)
± 40 μA max at V_0 = + 2.4 V

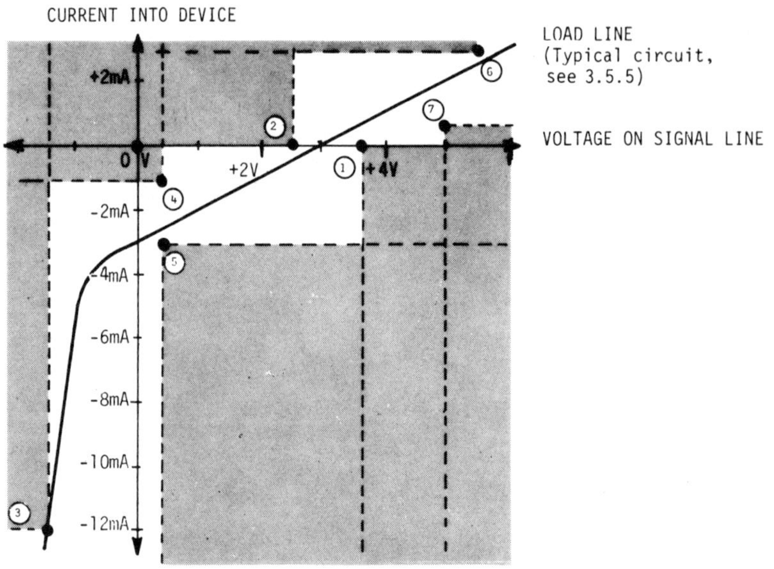

NOTE: The slope of the dc load line should, in general, correspond to a resistance not in excess of 3 kΩ.

Fig 15
DC Load Boundary Specification

Receiver: Input current
  −1.6 mA max at V_0 = +0.4V
 Input leakage current
  + 40 μA max at V_0 = +2.4 V
  + 1.0 mA max at V_0 = 5.25 V

V_{cc}: +5 V ± 5 percent

Only a single driver and receiver may be connected to each signal line in the typical configuration of Fig 16. Other configurations may exist in which this restriction does not hold provided the composite device load specifications of 3.5.3 are met.

3.6 Ground Requirements. The overall shield of the interconnecting cable shall be connected through one contact of the connector to frame (safety earth) to minimize susceptibility to and generation of external noise.

Warning: Devices should not be operated at significantly different frame potentials. The interface connection system may not be capable of handling excessive ground currents.

It is recommended that the ground returns of the individual control and status signal lines be connected to logic ground at the logic circuit driver or receiver to minimize cross-talk interference transients.

3.7 Cable Characteristics

3.7.1 Conductor Requirements. The maximum resistance for the cable conductors shall be, per meter of length:

(1) Each signal line (for example, DIO1, ATN), 0.14 Ω

(2) Each individual signal line ground return 0.14 Ω

(3) Common logic ground return 0.085 Ω

(4) Overall shield 0.0085 Ω

Fig 16
Typical Signal Line Input-Output Circuit

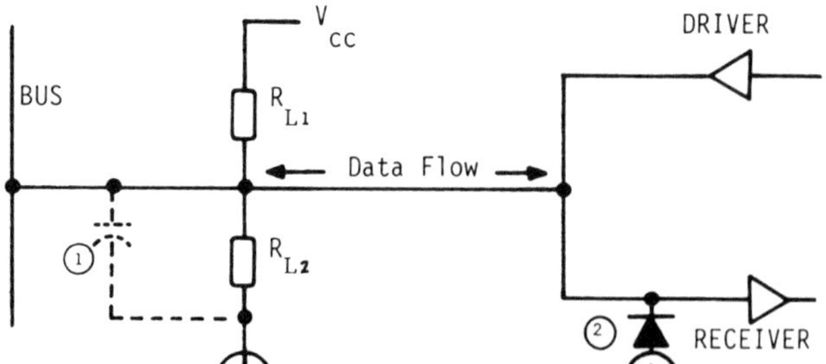

1 = Internal capacitance (cabling and transceivers) as defined by 3.5.4 and measured according to 3.7.2 for voltage values from 0 to 5 V.
2 = Typically contained within receiver component

3.7.2 Cable Construction. The cable should contain at least 24 conductors of which 16 shall be used for signal lines and the balance used for logic ground returns and overall shield.

The maximum capacitance measured (at 1 kHz) between any signal line and all other lines (signals, grounds, and shield) connected to ground shall be 150 pF per meter.

The shield shall contain a braid of 36 AWG wire or equivalent, with at least 85 percent coverage.

The cable shall be constructed to minimize the effects of cross talk between signal lines, the susceptibility of the signal lines to external noise, and the transmission of interface signals to the external environment.

(1) Each of the signal lines DAV, NRFD, NDAC, IFC, ATN, EOI, REN, and SRQ shall be twisted with one of the logic ground wires or isolated using an equivalent scheme to minimize cross talk.

(2) The cable shall contain an overall shield carried through the cable assembly and connectors at both ends to be returned to earth ground.

(3) A cable construction in which twisted pairs contained in the core of the cable and the individual DIO lines contained around the periphery of this core has been found satisfactory as has been the use of twisted pair conductors for all 16 signal lines where each signal line is twisted with an earth conductor.

(4) Alternately, any other internal cable construction which yields the same results may be used.

3.8 State Transition Timing Values. To ensure maximum possible compatibility among interconnected devices, Table 39 states the manda-

Table 39
Time Values

Time Value Identifier*	Function (applies to)	Description	Value
T_1	SH	settling time for multiline messages	$\geq 2\ \mu s$†
t_2	SH,AH,T,L,LE,TE	response to ATN	≤ 200 ns
T_3	AH	interface message accept time‡	> 0§
t_4	T,TE,L,LE,C,RL	response to IFC or REN false	$< 100\ \mu s$
t_5	PP	response to ATN $\wedge$ EOI	≤ 200 ns
T_6	C	parallel poll execution time	$\geq 2\ \mu s$
T_7	C	controller delay to allow current talker to see ATN message	≥ 500 ns
T_8	C	length of IFC or REN false	$> 100\ \mu s$
T_9	C	delay for EOI**	$\geq 1.5\ \mu s$††
T_{10}	C	delay for $\overline{\text{DAV}}$	$\geq 1.5\ \mu s$

*Time values specified by a lower case t indicate the maximum time allowed to make a state transition. Time values specified by an upper case T indicate the minimum time that a function shall remain in a state before exiting.

†If three-state drivers are used on the DIO, DAV, and EOI lines, T_1 may be:
(1) $\geq$ 1100 ns
(2) Or $\geq$ 700 ns if it is known that within the controller ATN is driven by a three-state driver (not recommended)
(3) Or $\geq$ 500 ns for all subsequent bytes following the first sent after each false transition of ATN (the first byte shall be sent in accordance with (1) or (2))
(4) Or $\geq$ 350 ns for all subsequent bytes following the first sent after each false transition of ATN under conditions specified in 5.2.3 and warning note.

‡Time required for interface functions to accept, not necessarily respond to, interface messages.

§ Implementation dependent.

**Delay required for EOI, NDAC, and NRFD signal lines to indicate valid states.

††$\geq$ 600 ns for three-state drivers.

tory time relationships between critical signal inputs and outputs to a specific device. The T_1, $T_6 - T_9$ time values allow for the normal propagation delays of the transmission path and the circuit delays within other devices. They are measured from the time the source output driver is seen to start its transition as viewed from its associated connector. It is further recommended that, for the minimum values of T_1, $T_6 - T_9$; the high-state driver voltage not be degraded, cable resistance and capacitance be kept as low as possible, and cross talk be kept at a minimum value.

4. Mechanical Specifications

4.1 Application. This section defines the mechanical specification for interface systems to be used in environments where:

(1) Physical distances between devices are limited

(2) Star or linear bus interconnection networks are useful

(3) Connector mounting space is limited

4.2 Connector Type. A quality connector of the rack and panel type with proven performance shall be used which has these minimum characteristics.

4.2.1 Electrical Considerations

Voltage rating; 200 V dc

Current rating; 5 A per contact

Contact resistance; $< 10\ m\Omega$

Contact material; gold over copper

Insulation resistance; $> 10\ G\Omega$ (diallyl-phthalate or equivalent)

4.2.2 Mechanical Considerations

Number of contacts; 24

Contact surfaces: self-wiping

Shell shape; trapezoidal polarization

Shell material; corrosion resistant material

Endurance; ≥ 1000 insertions

Wire termination diameter: accept at least $0.35\ mm^2$ (stranded)

Typical external dimensions (see 4.4 for additional dimensions);

A 46.8mm
B 15.5mm
C 2.16mm
D 4.29mm
E 3.33mm

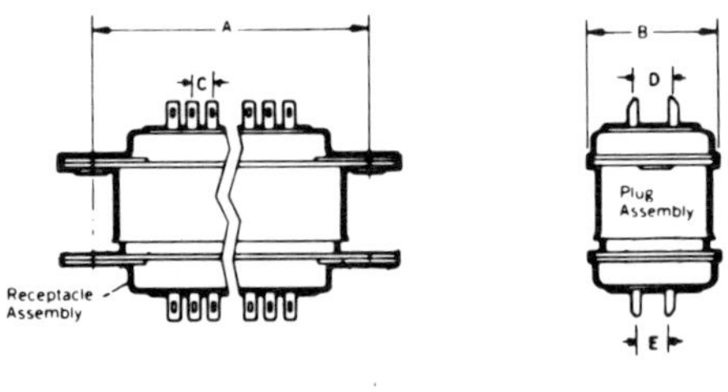

4.2.3 Environmental Considerations. Basic environmental performance relative to temperature and humidity criteria as tested in accordance with MIL STD 202E.[2]

4.3 Connector Contact Assignments. A contact assignment of the cable connector and the device connector shall be as shown below:

Contact	Signal Line	Contact	Signal Line
1	DIO 1	13	DIO 5
2	DIO 2	14	DIO 6
3	DIO 3	15	DIO 7
4	DIO 4	16	DIO 8
5	EOI (24)	17	REN (24)
6	DAV	18	Gnd, (6)
7	NRFD	19	Gnd, (7)
8	NDAC	20	Gnd, (8)
9	IFC	21	Gnd, (9)
10	SRQ	22	Gnd, (10)
11	ATN	23	Gnd, (11)
12	SHIELD	24	Gnd, LOGIC

NOTE: Gnd, (n) refers to the signal ground return of the referenced contact. EOI and REN return on contact 24.

4.4 Device Connector Mounting. Each device shall be provided with a receptacle type connector having these typical dimensions and with which the plug type connector shall mate. The two rows of twelve contacts each are centered within the trapezoidal shell. The connector mounting shall make provisions to accept the locking screws of the cable assembly.

The preferred orientation of the connector, as mounted on a device and viewed from the rear of the device in its normal operating position, is with contact 1 in the upper right-hand corner. The connector location should allow a minimum bend radius of 40 mm for cable clearance.

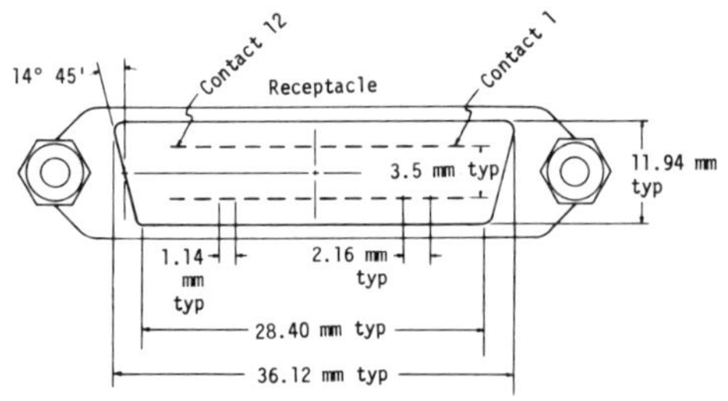

The connector may be mounted on either the outside or inside of the panel for which the

[2]The MICRORIBBON (Amphenol or Cinch Series 57) or CHAMP (AMP) connectors may be used for this application.

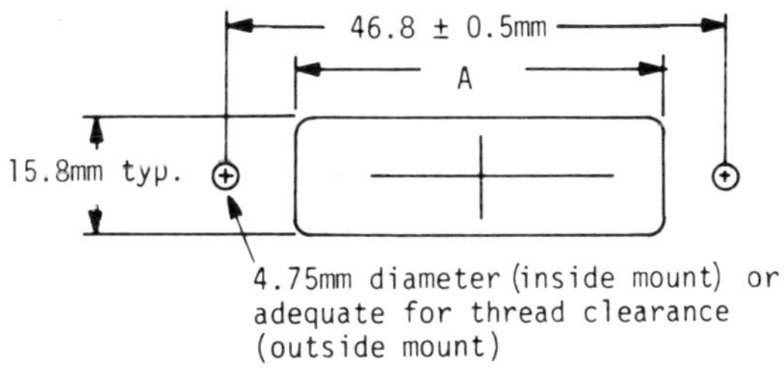

A Dimension	
36mm	outside mount
40.6mm	inside mount

Fig 17
Connector Panel Cutout

typical panel cutout dimensions are given in Fig 17.

The connector shall be attached to the device with one of the stud mount standoffs shown in Fig 18 as determined by the panel mounting method used.

4.5 Cable Assembly. The cable assembly shall be provided with both a plug and a receptacle connector type at each end of the cable. The preferred method of assembling the stacked connectors contains a rigid structure (assures a reliable and positive connection of multiple cable assemblies) as shown in Fig 19.

Each connector assembly shall be fitted with a pair of captive locking screws. Each lock screw shall conform to the mechanical dimension shown in Fig 20. A retraining ring[3], or equivalent, shall be used to retain the lock screw as a captive element.

It is recommended that each pair of connectors, assembled according to the first paragraph of 4.5, be partially enclosed within a suitable housing as shown in Fig 21.

Individual cable assemblies may be of any length up to 4 m.

[3] Such as TRUARC size 5133-12S or equivalent.

Fig 18
Mounting Dimensions

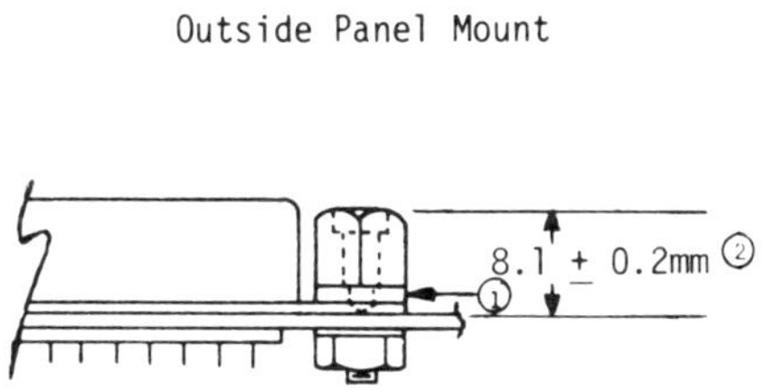

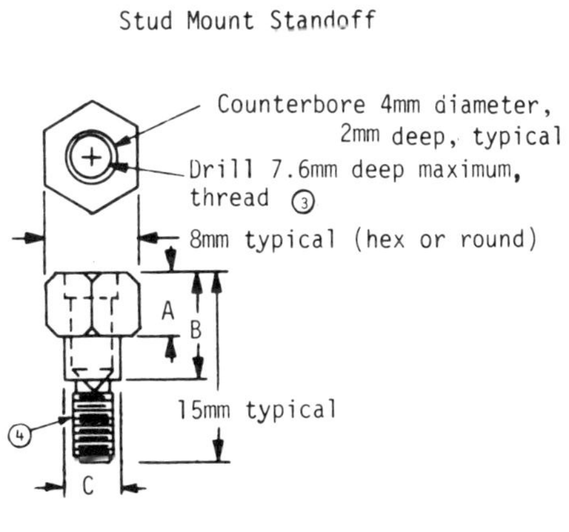

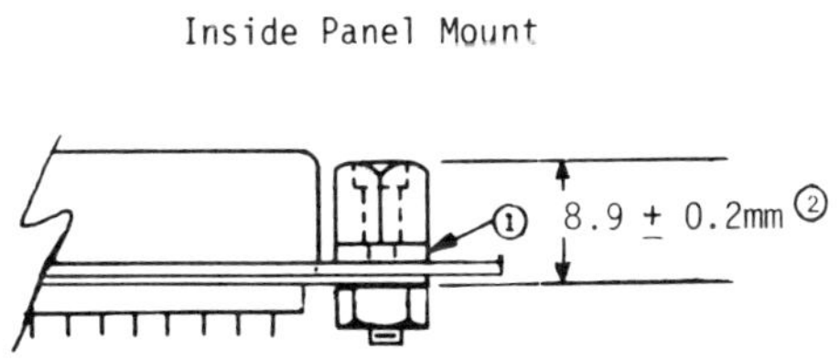

Standoff Dimensions	Outside Mount	Inside Mount
A	6.5mm	4.9mm
B	8.6mm	8.3mm
C	4.2mm	4.7mm

1 Split lock washer, 1.2 mm typical thickness
2 Typical dimension, determined by panel and lock washer thickness, connector housing
3 ISO metric thread M3.5 × 0.6 or equivalent Optimum Metric Fastener System (OMFS) thread 3.5PO.6
4 Thread size designers choice (does not affect compatibility)

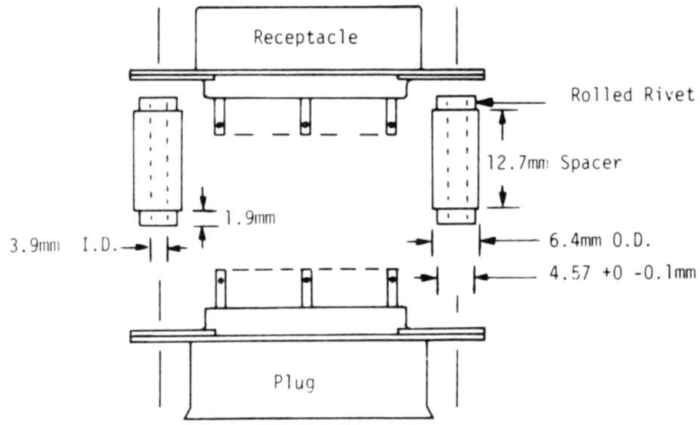

Fig 19
Stacked Connector Assembly

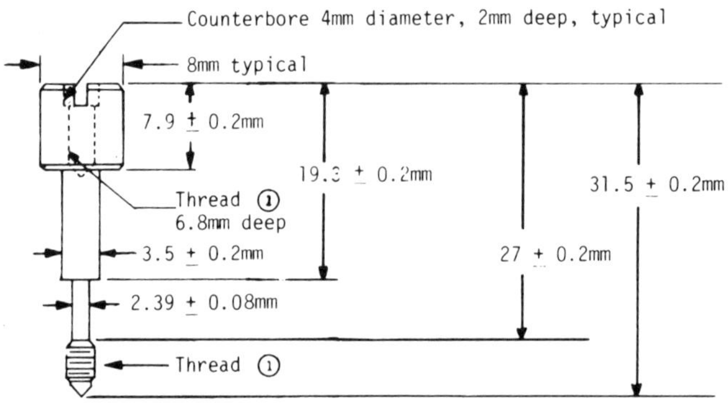

1 = ISO metric thread M3.5×0.6 or equivalent O.M.F.S. thread 3.5PO.6

Fig 20
Lock Screw

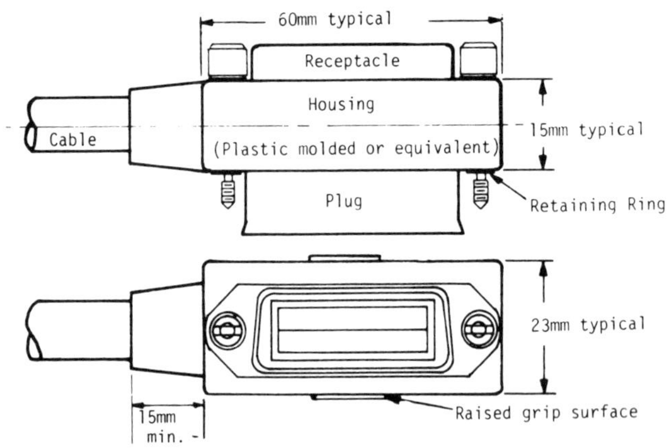

Fig 21
Cable Connector Housing

5. System Applications and Guidelines for the Designer

5.1 System Compatibility. This interface system offers a wide range of capability from which to choose the appropriate interface functions to fit different applications. Within most interface functions a number of options are available. In addition, the designer has freedom to select all the device dependent capabilities contained within the device functions.

It is the responsibility of the designer to define the complete capability of a device (interface system choices and related device dependent interactions) so that the end user of the device can efficiently interface and program the device for appropriate system applications.

Selection of a minimum set of interface functions from Section 2 leads to the following minimum set of signal lines in order to be system compatible:

(1) DIO 1-7

(2) DAV, NRFD, NDAC

(3) IFC and ATN (unnecessary in systems without a controller)

In order to provide system compatibility a designer shall not introduce new interface functions beyond those defined in Section 2.

5.2 Data Rate Consideration. Designers of devices intended to communicate over the interface system bus are advised to consider the relationships between various levels of system performance and the specific device circuits used to provide these different levels of performance. The following statements are intended as guidance.

5.2.1 The interface bus will operate at distances up to 20 m, at a maximum of 250 000 bytes per second, with an equivalent standard load for each 2 m of cable using 48 mA open collector drivers.

5.2.2 The interface bus will also operate at distances up to 20 m at a maximum of 500 000 bytes per second, with an equivalent standard load for each 2 m of cable using 48 mA three-state drivers.

5.2.3 Higher Speed Operation. To achieve the maximum possible data transfer rate (nominally up to 1 000 000 bytes per second) within a system, the following guidelines should be observed:

(1) All devices expected to talk at higher rates should use a minimum T_1 value of 350 ns

(2) All devices expected to operate at higher rates should use 48 mA three-state drivers

(3) The device capacitance on each lead (REN and IFC excepted) should be less than 50 pF per device. In a system configuration the total device capacitance should be no more than 50 pF for each equivalent resistive load in the system.

(4) All the devices in the system should be powered on

(5) Interconnecting cable links should be as short as possible up to a maximum of 15 m total length per system with at least one equivalent load for each meter of cable

WARNING: Anytime a device following condition (1) is placed in a system, *even if higher speed operation is not intended*, there may be data transfer errors if conditions (2) through (5) are not met for that system.

NOTES:

(1) Devices with a T_1 value of less than 500 ns, device capacitance of 50 pF, or having multiple resistive loads shall be so marked, as acceptable variants. Multiple resistive powered loads, beyond one per signal line per device, can be added up to a maximum of 15 loads total per signal line per system. Multiple loads can be used, with caution, to improve the device to cable-length ratio (up to 15 m max).

(2) Actual maximum data rates may be more a function of device dependent time delays than the cable characteristics or timing values of 3.8.

(3) See 3.5.4 for warning of adverse effects of variable capacitive loading.

(4) Use of data byte buffer store within the device may be advantageous.

5.3 Device Capabilities

5.3.1 Busy Function. In system operation it is useful to either program a device or initiate some operation within a device and then proceed to communicate with other devices (while the first device is busy carrying out the required task). The busy (operation being completed) function is a device state and not an interface state. In order to permit interface bus communication indepenent of the busy condition of a device, three possible methods are available:

(1) SRQ and serial poll

(2) Parallel poll

(3) NFRD hold

Both the serial poll and parallel poll methods are described in Section 2.

5.3.2 NFRD Hold. The NRFD signal line may be gated to incorporate the busy function.

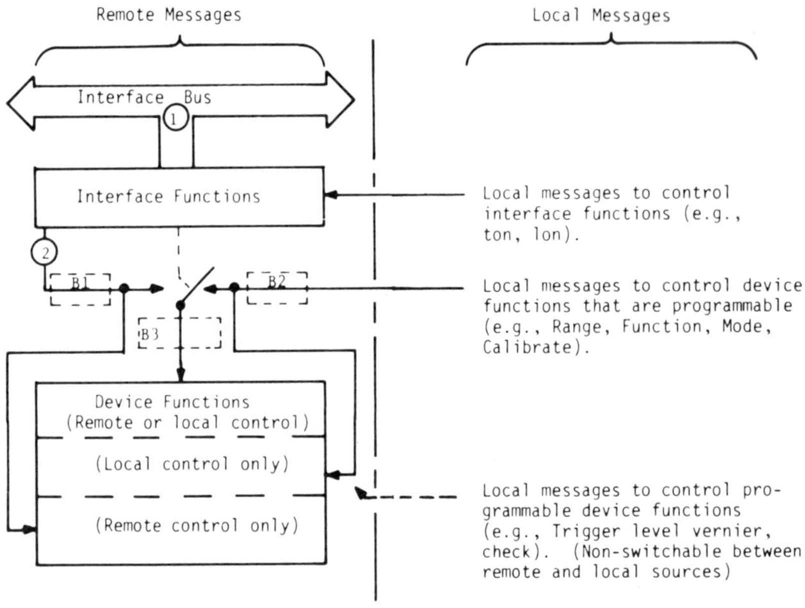

1 = Remote messages (for example, ATN, MLA interface messages, DAB device messages)
2 = Remote control of programmable device functions (for example, range, function) and the execution of device functions (for example, clearing, triggering)
B = Possible locations of buffer storage elements. Normally B1 and B2 are used in combination or B3 is used alone.

Fig 22
Remote — Local Message Paths

In so doing, the NRFD signal (or RFD message) changes its definition to include more than the normal "ready for the next data byte" meaning. The internal busy signal is gated to the NRFD signal line through the AH function. In this manner the device may be unaddressed as listener during a "busy cycle" and the interface bus may be used for other purposes. When readdressed as a listener, the device will indicate its internal busy status to the interface. The device indicates "busy" by setting NRFD to 1 and indicates "operation complete" by setting NRFD to Ø.

Caution: If NRFD hold is used for the busy function where a device may not recover or may never reach the non-busy condition, then another listen address (always accessible) should be available to clear the potential hang-up condition.

5.3.3 RL Applications. The designer is free to implement, within a device, whatever programmable device functions are appropriate for a particular device application(s). The designer is not free to remotely program the local control functions which interact directly with the interface functions as specified throughout Section 2.

To implement a programmable device capable of being either remotely or locally controlled may require the switching of some or all of the typical controls illustrated in Fig 22. This figure is not meant to imply a comprehensive set of switching techniques, switching locations, or switched message contents.

5.4 AND and OR Logic Operations. The message sent by one interface function is not necessarily the same message received by another interface function, (irrespective of time differences due to transmission characteristics of the signal lines) in the case of three messages as used in the SH, AH, and SR interface functions:

(1) The RFD (or DAC) message received (by an SH function) shall be the logic AND of all RFD (or DAC) messages sent (by all AH functions)

(2) The SRQ message received (by a C) shall be the logic OR of all SRQ messages sent (by the SR functions)

NOTE: The DAV message received (by all AH functions) shall be the DAV message sent (by one and only one SH function).

5.4.1 RFD and DAC Messages. The RFD (or DAC) message sent true (or false) by an AH function is performed by setting the NRFD (or NDAC) signal line to ∅ (high) or driving the NRFD (or NDAC) signal line to 1 (low), respectively.

The RFD (of DAC) message received by an SH function is received true when the state of the signal line is ∅ (high) which means that all RFD (or DAC) messages sent are sent passive true.

The RFD (or DAC) message received by an SH function is received false when the state of the signal line is 1 (low) which means that one or more RFD (or DAC) messages sent are sent false.

The logical equivalent of these conditions is illustrated below.

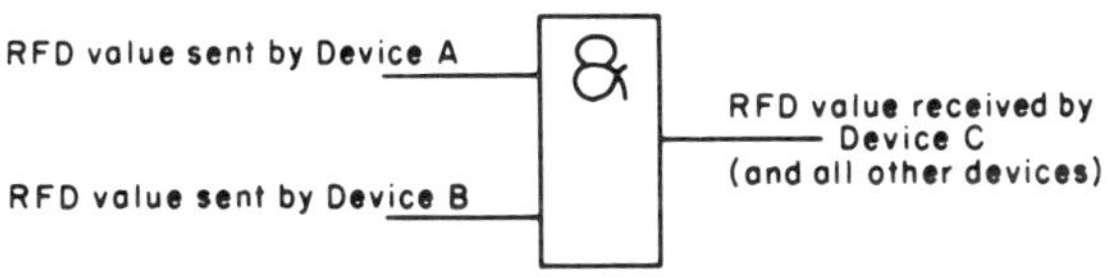

5.4.2 SRQ Message. The SRQ message sent true, or false, by an SR function is performed by driving the SRQ signal line to 1 (low) or setting the SRQ signal line to ∅ (high), respectively.

The SRQ message received by the C function is received true when the state of the bus signal line is 1 (low) which means that one or more SRQ functions have sent the SRQ message true.

The SRQ message received by a controller function is received false when the state of the bus signal line is ∅ (high) which means that all SRQ functions have sent the SRQ message passive false.

The logical equivalent of these conditions is illustrated below.

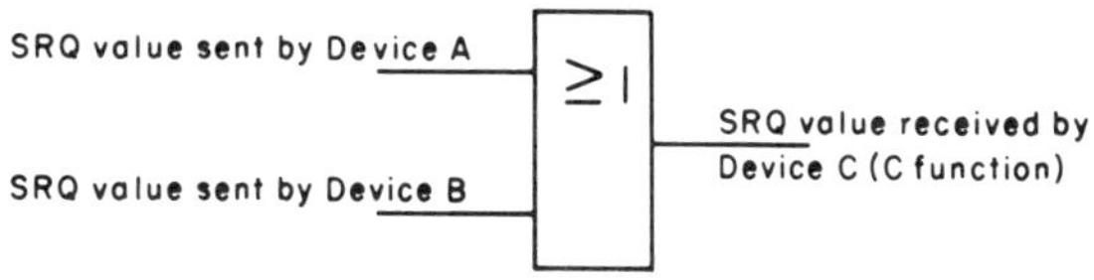

5.4.3 Circuit Implementations. A typical circuit configuration with which the AND and OR functions on the respective bus signal lines can be perfromed is that represented in 3.5.5, Fig 16. The driver element shall be a bi-state (open collector) driver as represented in Fig 23.

NOTE: Whether or not invertors are used to convert the internal representation of the RFD (or DAC) message into the actual message sent on the bus signal lines depends on the internal assertion definition for true and false with respect to the high or low voltage levels used internal to the device. This matter is left to the designer.

Typical signals presented to the NRFD (or SRQ) interface bus signal lines by devices *A* and *B* as described in 5.4.1 and 5.4.2 may be represented as shown in Fig 24. Only the composite signal line waveform as received at device *C* exists on the bus. The signal levels shown for devices *A* and *B* exist only within the devices' drivers and not on the bus signal line.

5.5 Address Assignment. Normally, a device will be assigned a single talk and single listen

Fig 23
Bi-State Signal Line Logic
(Open Collector Drivers)

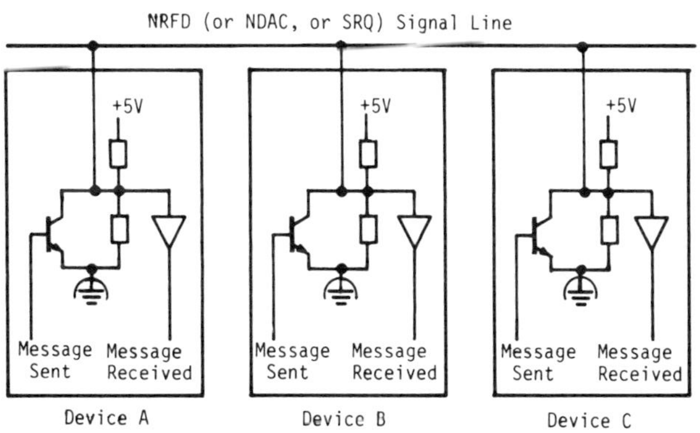

address to perform the essential tasks. It may be useful to design a device with multiple talk (or listen) addresses to facilitate system requirements. A device could be assigned two talk addresses (for example, one to output raw data, the other to output processed data). Care should be given to minimize the use of such multiple addresses as later system configurations may be restricted due to excessive use of primary addressing capability.

5.6 Typical Combinations of Interface Functions. The designer is free to select the particular interface functions required to meet specific device applications. The selection of certain interface functions requires the inclusion of other interface functions as defined throughout the allowable subset clauses of Section 2.

Device	Typical Interface Functions Used
Signal generator (Only able to listen)	AH, L, RL, DT
Tape reader (Only able to talk)	SH, AH, T
Digital voltmeter (Able to talk and listen)	SH, AH, T, L, SR, RL, PP, DC, DT
Calculator (Able to talk, listen, and control)	SH, AH, T, L, C

The list above represents typical combinations of interface functions and does not imply that these are the only combinations possible or useful.

Fig 24
Signal Line Logic and Timing Relationships — (a) NRFD (or NDAC) Signal Line; (b) SRQ Signal Line

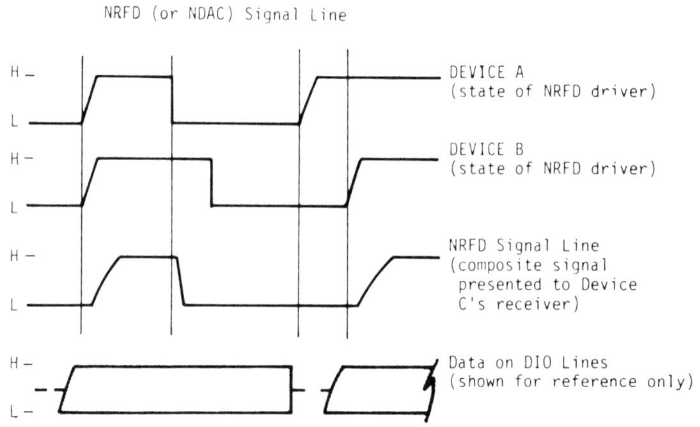

(a)

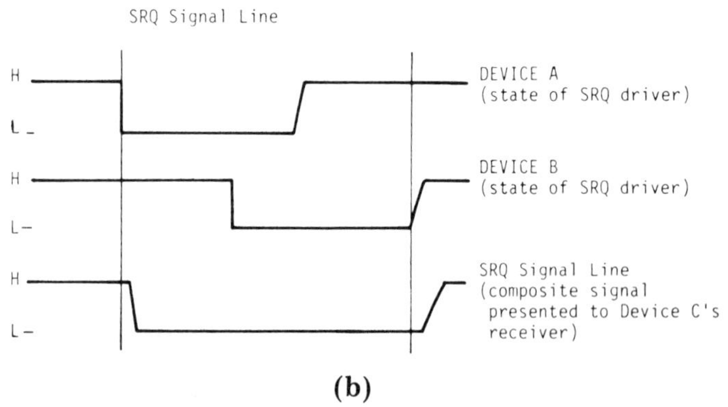

(b)

6. System Requirements and Guidelines for the User

6.1 System Compatibility. Devices designed to this interface system may have a wide range of capability relative to their ability to communicate over the interface. This standard does not cover the operational characteristics of devices, only the mechanical, electrical, and functional capabilities of the interface system.

The burden of responsibility for system compatibility at the operational level is on the user. The user shall be familiar with all device characteristics interacting with the interface system (for example, device dependent program codes, output data format and codes, etc).

6.2 System Installation Requirements. This includes system configuration restrictions.

6.2.1 Maximum Number of Devices. The maximum number of devices that can be connected together to form one interface system is 15.

6.2.2 Minimum System Configurations. An interface system shall contain one or more devices containing at least one T function, one L function, and one C function.

If all the T functions include the use of the ton message (talker types T1, T3, T5, T7, TE1, TE3, TE5, or TE7), and all the L functions include the lon message (listener types L1, L3, LE1 or LE3), a system may be operated without a C function when the ton and lon messages are true. The lon and ton messages are normally provided by local switches.

6.2.3 System Controllers. All system configurations containing more than one controller shall satisfy the following conditions:

(1) There shall not be more than one C function in a system that is in the SACS (system control active state)

(2) Every controller in the system shall be able to pass and receive control of the interface

6.2.4 Devices Powered Off and On. A system will operate without adversely affecting normal data transfer with at least two thirds of the devices powered on. A system will operate correctly with any number of devices powered off provided all those devices powered off do not degrade the specified high state condition, that is, that the voltage on each signal line with all its output drivers passive false should exceed +2.5 V with respect to the logic ground at each device.

Unless special precautions are taken (that is, use of special driver circuits beyond the scope of this standard) powering a device to on while the system is running may cause faulty operation.

6.3 Address Assignment

6.3.1 Talk Addresses. A device that contains a T function or a TE function may be assigned any value for bits T1 through T5 of its MTA (my talk address) message code other than:

T5	T4	T3	T2	T1
1	1	1	1	1

This code, defined as UNT, is provided as a systems convenience, for the controller, to return all devices to the talker idle state.

Two or more T functions (whether within the same or separate devices) shall not be assigned the same value for bits T1 through T5 of their MTA codes.

A device that contains both a T and L function may be assigned a talk address such that T1 through T5 of its MTA code equals L1 through L5 of its MLA code.

A TE interface function shall not be assigned the same value for bits T1 through T5 of its MTA code as that assigned to a T function.

6.3.2 Listen Addresses. A device that contains an L function or an LE function may be assigned any value for bits L1 through L5 of its MLA (my listen address) code other than:

L5	L4	L3	L2	L1
1	1	1	1	1

Two or more L functions (usually within separate devices) may be assigned the same value for bits L1 through L5 of their MLA codes.

A device that contains both an L and T function may be assigned a listen address such that L1 through L5 of its MLA code equals T1 through T5 of its MTA code.

6.3.3 Secondary Addresses. A device that contains a TE function or LE function may be assigned any value for bits S1 through S5 of its MSA (my secondary address) code other than:

S5	S4	S3	S2	S1
1	1	1	1	1

Two or more TE functions (whether within the same or separate devices) shall not be assigned the same value for bits T1 through T5 of their MTA codes and bits S1 through S5 of their MSA codes.

Two or more LE functions (usually within separate devices) may be assigned the same value for both bits L1 through L5 of their MLA codes and bits S1 through S5 of their MSA codes.

A device that contains both a TE and LE function may be assigned a listen address such that L1 through L5 of its MLA code equals T1 through T5 of its MTA code and both functions may utilize the same secondary address.

6.4 Cabling Restrictions

6.4.1 Maximum Cable Length. The maximum length of cable that shall be used to connect together a group of devices within one bus system is:

(1) 2 m times the number of devices

(2) Or 20 m, whichever is less

6.4.2 Distribution of Maximum Cable Lengths. The maximum length of cable as defined in 6.4.1 may be distributed among the devices in a system in any manner deemed suitable by the user. Caution should be taken if individual cable length exceeds 4 m.

6.4.3 Cabling Configurations. Cables may be interconnected in any manner deemed suitable by the user (that is, star, linear, or combinations thereof).

Devices should not be operated at significantly different frame potentials as the system may not be capable of handling excessive ground currents.

6.5 Operational Sequence Guidelines. Most interface communication tasks require a sequence of coded messages to be sent over the interface. Although the specification of operational sequences is beyond the scope of this standard, several sequences are recommended for typical tasks. Many other sequences might be found useful.

NOTE: Caution should be observed by the system user to assure that exit conditions from a given sequence leaves devices in an acceptable state. Adequate device documentation facilitates this process.

6.5.1 Data Transfer

ATN		
[1	UNL]	Inhibits all current listeners (can be omitted if not required)
1	$(LAD)_1$	Each address sent enables a specific device to receive future data bytes
[	⋮	More than one address may be sent if multiple listeners are desired
1	$(LAD)_n$]	
1	(TAD)	The address sent enables a specific device to send data as soon as ATN becomes Ø
Ø	$(DAB)_1$	Sent by the currently enabled talker to all currently enabled listeners
[	⋮	Bytes may be sent until the controller again sets ATN to 1 to repeat the sequence. If the talker is sending a specific-length record, it may optionally set EOI=1 concurrent with the last byte
Ø	$(DAB)_n$]	
	⋮	

NOTE: (LAD) represents a listen address of a specific device.
(TAD) represents a talk address of a specific device.
(DAB) represents any data byte.
[ ] indicates optional segments of a sequence.
() indicates messages not uniquely defined in this standard.

6.5.2 Serial Poll (issued by controller usually whenever SRQ=1 on the interface)

ATN		
1	UNL	Prevents other devices from listening to status sent (controller continues to listen without being addressed)
1	SPE	Puts interface into serial poll mode during which all devices send status instead of data when enabled
1	$(TAD)_n$	Enables a specific device to send status. Within this loop, devices should be sequentially enabled
Ø	(SBN) or (SBA)	Status byte sent by enabled device. If SBN was sent, loop should be repeated. If SBA was sent, the enabled device is identified as having sent SRQ over the interface and will automatically remove it
	or	
1	SPD	Removes the interface from serial poll mode
1	[UNT]	Disables last talker from sending data if ATN is set Ø

NOTE: (TAD) represents a talk address of a specific device.
(SBN) represents a status byte sent by a device in which a request for service is not indicated (bit 7=Ø) ($SBN = STB \wedge \overline{RQS}$).
(SBA) represents a status byte sent by a device in which a request for service is indicated (bit 7=1) ($SBA = STB \wedge RQS$).

6.5.3 Control Passing

ATN		
1	(TAD)	The address sent should be that of the device to which control is being passed
1	TCT	Notifies addressed device to take over control of the interface
1		New controller-in-charge at this time

NOTE: (TAD) represents a talk address of a specific device.

6.5.4 Parallel Poll

6.5.4.1 *Parallel Poll Configure*

ATN		
1	(LAD)	Addresses a particular device for which a parallel response coding is to be assigned
1	PPC	Enables the addressed listener to be configured
1	PPE	Bit 4 specifies the sense of the poll response, bits 1. . .3 specify, in binary code, the DIO line on which the poll response is to be given
1	UNL	End of the configuration routine

NOTE: (1) (LAD) represents a listen address of a specific device.
(2) The PPE command can be cleared by a PPD command.
(3) The configuration can be cleared by a PPU command.

6.5.4.2 *Parallel Poll Response*

ATN	IDY	
1	1	Whenever the bus is in this state, predetermined devices will each place their requests on a specific DIO line. If more than one device is sharing a DIO line, the line value can indicate either an ORing or an ANDing of requests depending on commands previously sent to the devices instructing them to use the Ø or 1 value to request service.

6.5.5 Placing Devices in Forced Remote Control

ATN	REN		
1	1	LLO	Disables all devices' "return-to-local" button
1	1	$(LAD)_1$	Each address sent places the addressed device into remote state, disabling all local controls
1	1	.	
1	1	.	
		.	
1	1	$(LAD)_n$	

NOTE: (LAD) represents a listen address of a specific device. (Devices will all revert back to local state as a group at any time a $\emptyset$ value of REN is placed on the interface.)

6.5.6 Sending Interface Clear. While the IFC message is being sent only the DCL, LLO, PPU, and REN universal commands will be recognized.

Appendixes

(These Appendixes are not a part of IEEE Std 488-1978, Digital Interface for Progammable Instrumentation, but are included solely for purposes of information.)

Appendix A
Typical Instrument System

The typical system shown in Fig A1 illustrates the capability of the interface system to handle a variety of instrumentation system needs. Two possible event sequences, to accomplish specific measurement tasks using the interface system, are included as examples.

A1. Event Sequence 1 (Device Dependent Data Returned to Processor)

Processor programs instruments and initiates measurements; resulting basic data is returned to processor.

(1) Processor initializes the interface system by sending the IFC message true.

(2) Processor causes all devices to set their internal conditions to be a predefined state by sending the DCL message true.

(3) Processor sends the listen address of the dc power supply followed by program data for that device.

(4) Processor sends the unlisten command, then the listen address for the next device, followed by program data for it.

(5) Event (4) is repeated until each device of interest for this specific test has been addressed and programmed, then the unlisten command is sent.

(6) Processor sends listen address of selected measurement device (for example, the digital frequency meter), then that program code required to initiate a measurement.

(7) Processor sends unlisten command, addresses itself to listen, then sends talk address of the measurement device.

(8) Upon completion of its internal measurement cycle, the digital frequency meter sends (talks) its measurement results (device dependent data) to the addressed listener, the processor.

Fig A1
Typical System Showing Capability of Interface System to Handle Variety of Instrumentation System Needs

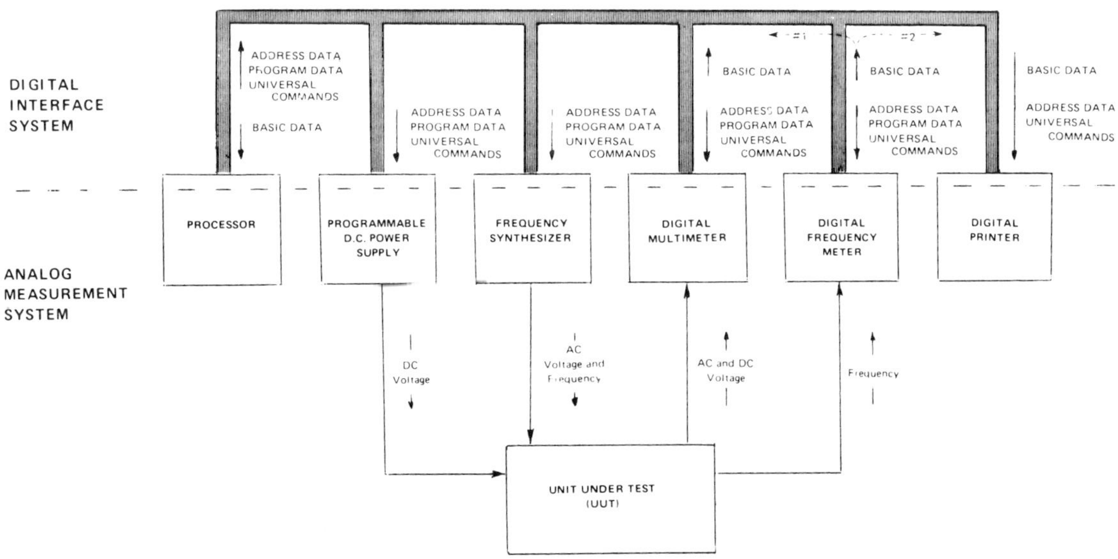

A2. Event Sequence 2 (Device Dependent Data Directed to Digital Printer)

Processor programs instruments and initiates measurements; resulting device dependent data is returned to another device.

(1) — (6) Identical to Event Sequence 1.

(7) Processor sends unlisten command, then the listen address of the digital recorder, followed by the talk address of the measurement device.

(8) Upon completion of its measurement, the measurement device again sends its resulting device dependent data to the addressed listener, the digital recorder.

NOTE: If the processor were to address both the digital recorder and itself, the resulting device dependent data would be accepted by both devices, even though the two may have vastly different rates at which data can be accepted.

Appendix B
Handshake Process Timing Sequence

B1. General Comments

Each data byte transferred by the interface system uses the handshake process to exchange data between source and acceptor. Typically, the source is a talker and the acceptor a listener.

Fig B1 illustrates the handshake process by indicating the actual waveforms on the DAV, NRFD, and NDAC signal lines. The NRFD and NDAC signals each represent composite waveforms resulting from two or more listeners accepting the same data byte at slightly different times due to variations in the transmission path length and different response rates (delays) to accept and process the data byte.

Fig B2 represents the same sequence of events, in flow chart form, to transfer a data byte between source and acceptor.

The annotation numbers on fhe flow chart and the timing sequence diagram refer to the same event on the list of events.

B2. List of Events for Handshake Process

(1) — Source initializes DAV to high (H) (data not valid).

(2) — Acceptors initialize NRFD to low (L) (none are ready for data), and set NDAC to low (L) (none have accepted the data).

(3) t_{-2} Source checks for error condition (both NRFD and NDAC high), then sets data byte on DIO lines.

Fig B1
Signal Line Timing Sequence for One Talker and Multiple Listeners
Using Handshake Process
(See Fig B2 and List of Events) $H \geq +2.0$ V; $L \leq +0.8$ V

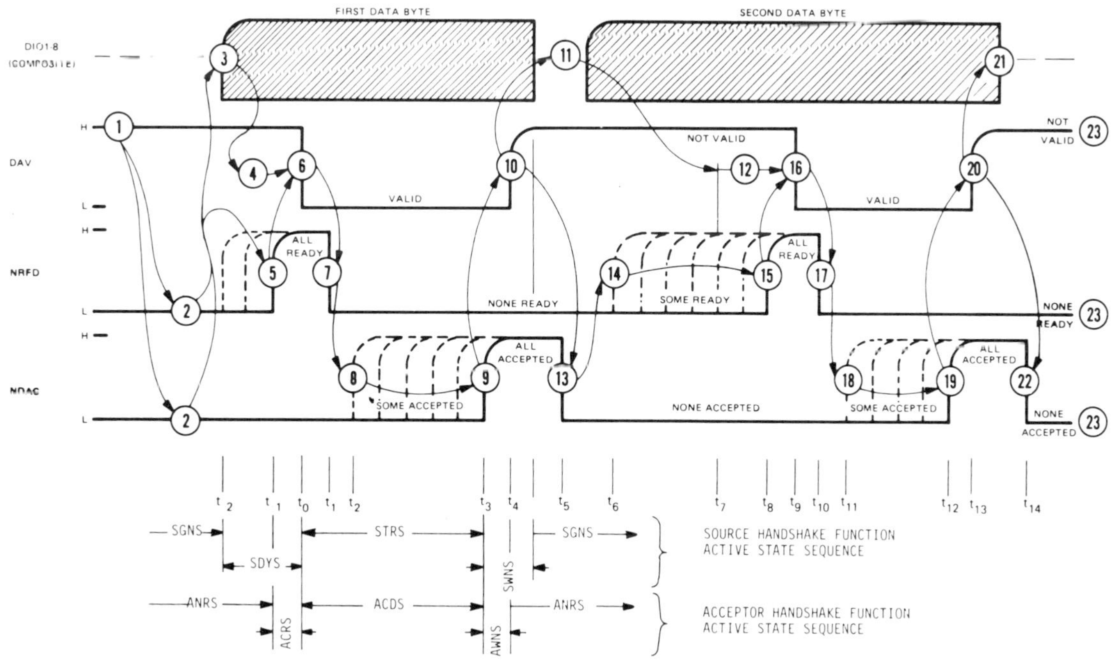

(4) $t_{-2} \rightarrow t_0$ Source delays to allow data to settle on DIO lines.

(5) t_{-1} Acceptors have all indicated readiness to accept first data byte; NRFD lines goes high.

(6) t_0 Source, upon sensing NRFD high, sets DAV low to indicate that data on DIO lines is settled and valid.

(7) t_1 First acceptor sets NRFD low to indicate that it is no longer ready, then accepts the data. Other acceptors follow at their own rates.

(8) t_2 First acceptor sets NDAC high to indicate that it has accepted the data. (NDAC remains low due to other acceptors driving

Fig B2
Logical Flow of Events for Source and Acceptor When Transferring Data Using Handshake Process
(See List of Events) (This flow diagram is not intended to represent the only method of implementing an acceptor handshake. See 2.4.5, paragraph three)

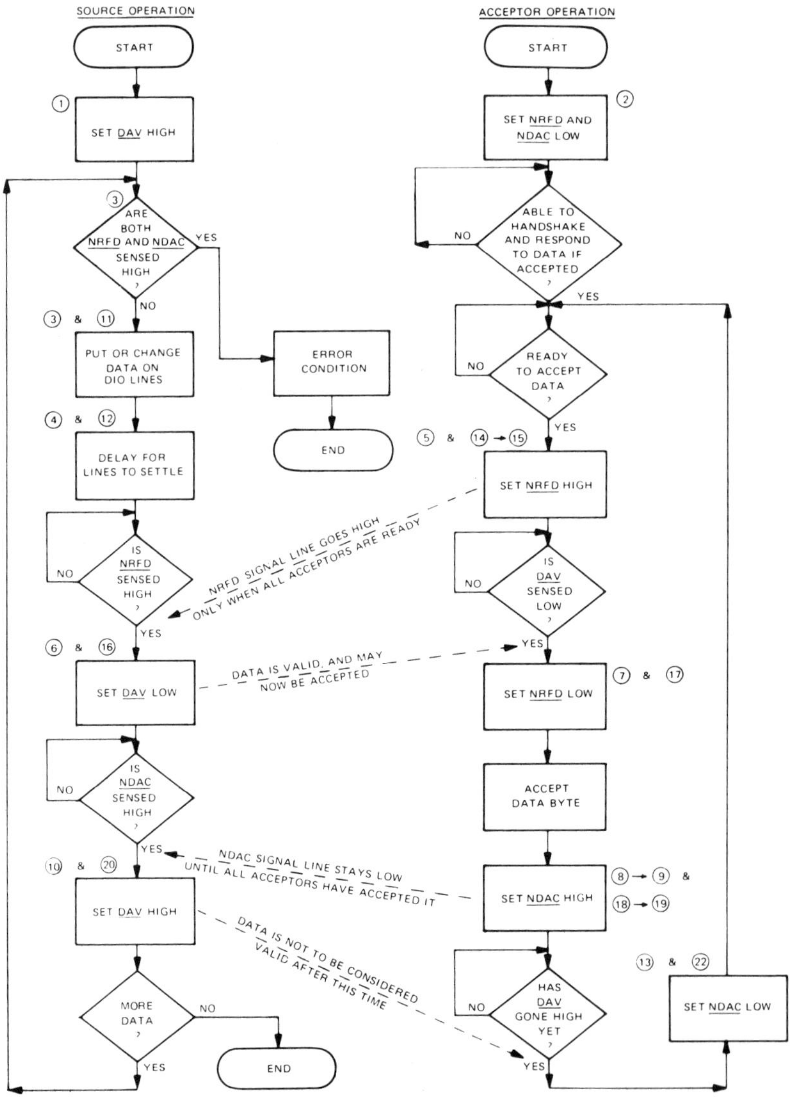

NDAC low).

(9) t_3 Last acceptor sets NDAC high to indicate that it has accepted the data; all have now accepted and the NDAC line goes high.

(10) t_4 Source, having sensed that NDAC is high, sets DAV high. This indicates to the acceptors that data on the DIO lines must now be considered not valid.

(11) t_4—t_7 Source changes data on the DIO lines.

(12) t_7—t_9 Source delays to allow data to settle on DIO lines.

(13) t_5 Acceptors, upon sensing DAV high (at 10) set NDAC low in preparation for next cycle. NDAC line goes low as the first acceptor sets the line low.

(14) t_6 First acceptor indicates that it is ready for the next data byte by setting NRFD high. (NRFD remains low due to other acceptors driving NRFD low).

(15) t_8 Last acceptor indicates that it is ready for the next data byte by setting NRFD high; NRFD signal line goes high.

(16) t_9 Source, upon sensing NRFD high, sets DAV low to indicate that data on DIO lines is settled and valid.

(17) t_{10} First acceptor sets NRFD low to indicate that it is no longer ready, then accepts the data.

(18) t_{11} First acceptor sets NDAC high to indicate that it has accepted the data [as in (8)].

(19) t_{12} Last acceptor sets NDAC high to indicate that it has accepted the data [as in (9)].

(20) t_{13} Source, having sensed that NDAC is high, sets DAV high [as in (10)].

(21) — Source removes data byte from DIO signal lines after setting DAV high.

(22) t_{14} Acceptors, upon sensing DAV high, set NDAC low in preparation for next cycle.

(23) — Note that all three handshake lines are at their initialized states [as in (1) and (2)].

Appendix C
Interface Function Allowable Subsets

C1. General Comments

Sections 5 and 6 of this standard point out device designer responsibility to identify and device user responsibility to be familiar with the interface capabilities of each device containing IEEE Std 488-1978 interface functions. It is therefore recommended that each device be marked with an explicit code which, in effect, identifies those interface functions and subsets implemented within that device.

The reader is further reminded that full operational systems require detailed knowledge of the device dependent characteristics of each device in a system. These specifications are beyond the scope of this standard.

C2. Capability Identification Codes

It is recommended that an IEEE Std 488-1978 capability code be placed below or near the interface connector on each device to identify the complete set of interface functions contained within that device. In this standard each interface function and allowable subset thereof has an equivalent alphanumeric code to identify that particular capability. All such interface function capability codes may be expressed in a concise alphanumeric string and marked on the exterior of the device to facilitate user system assembly.

For example, a device with the basic talker function, the ability to send status bytes, the basic listener function, a listen only mode switch, service request capability, remote local capability without local lockout, manual configuration of the parallel poll capability, complete device clear capability, no capability for device trigger, and no controller capability would be identified with the following code:

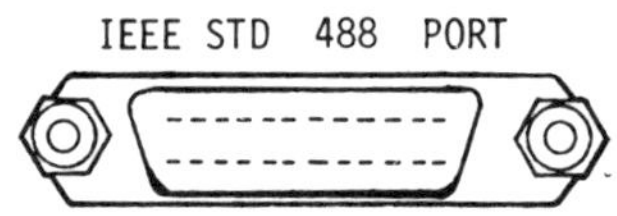

SH1, AH1, T2, L1, SR1, RL2, PP2, DC1, DTØ, CØ, E1

The code identifies the eight specific interface functions implemented. In addition, the type of electrical interface contained within the device is specified. The notation E1 is used to identify open collector drivers and E2 is used to identify three-state drivers.

The device designer can place any further device capability information, useful for system configuration, at the appropriate places on the physical equipment and in the relevant documentation for that equipment.

C3. SH Function Allowable Subsets

Identification	Description	States Omitted	Other Requirements	Other Function Subsets Required
SH0	no capability	all	none	none
SH1	complete capability	none	none	T1-T8, TE1-TE8, or C5-C28

C4. AH Function Allowable Subsets

Identification	Description	States Omitted	Other Requirements	Other Function Subsets Required
AH0	no capability	all	none	none
AH1	complete capability	none	none	none

C5. T Function Allowable Subsets

Identification	Description: Capabilities				States Omitted	Other Requirements	Other Function Subsets Required
	Basic Talker	Serial Poll	Talk Only Mode	Unaddress If MLA			
T0	N	N	N	N	all	none	none
T1	Y	Y	Y	N	none	omit [MLA $\wedge$ (ACDS)]	SH1 and AH1
T2	Y	Y	N	N	none	omit [MLA $\wedge$ (ACDS)] ton always false	SH1 and AH1
T3	Y	N	Y	N	SPIS, SPMS, SPAS	omit [MLA $\wedge$ (ACDS)]	SH1 and AH1
T4	Y	N	N	N	SPIS, SPMS, SPAS	omit [MLA $\wedge$ (ACDS)] ton always false	SH1 and AH1
T5	Y	Y	Y	Y	none	include [MLA $\wedge$ (ACDS)]	SH1 and L1-L4 or LE1-LE4
T6	Y	Y	N	Y	none	include [MLA $\wedge$ (ACDS)] ton always false	SH1 and L1-L4 or LE1-LE4
T7	Y	N	Y	Y	SPIS, SPMS, SPAS	include [MLA $\wedge$ (ACDS)]	SH1 and L1-L4 or LE1-LE4
T8	Y	N	N	Y	SPIS, SPMS, SPAS	include [MLA $\wedge$ (ACDS)] ton always false	SH1 and L1-L4 or LE1-LE4

C6. T Function (with Address Extension) Allowable Subsets

Identi-fication	Description				States Omitted	Other Requirements	Other Function Subsets Required
	Capabilities						
	Basic Extended Talker	Serial Poll	Talk Only Mode	Unaddress If MSA ∧ (LPAS)			
TE0	N	N	N	N	all	none	none
TE1	Y	Y	Y	N	none	omit [MSA ∧ (LPAS) ∧ (ACDS)]	SH1 and AH1
TE2	Y	Y	N	N	none	omit [MSA ∧ (LPAS) ∧ (ACDS)] ton always false	SH1 and AH1
TE3	Y	N	Y	N	SPIS, SPMS, SPAS	omit [MSA ∧ (LPAS) ∧ (ACDS)]	SH1 and AH1
TE4	Y	N	N	N	SPIS, SPMS, SPAS	omit [MSA ∧ (LPAS) ∧ (ACDS)] ton always false	SH1 and AH1
TE5	Y	Y	Y	Y	none	include [MSA ∧ (LPAS) ∧ (ACDS)]	SH1 and L1-L4 or LE1-LE4
TE6	Y	Y	N	Y	none	include [MSA ∧ (LPAS) ∧ (ACDS)] ton always false	SH1 and L1-L4 or LE1-LE4
TE7	Y	N	Y	Y	SPIS, SPMS, SPAS	include [MSA ∧ (LPAS) ∧ (ACDS)]	SH1 and L1-L4 or LE1-LE4
TE8	Y	N	N	Y	SPIS, SPMS, SPAS	include [MSA ∧ (LPAS) ∧ (ACDS)] ton always false	SH1 and L1-L4 or LE1-LE4

C7. L Function Allowable Subsets

Identi-fication	Description			States Omitted	Other Requirements	Other Function Subsets Required
	Capabilities					
	Basic Listener	Listen Only Mode	Unaddress If MTA			
L0	N	N	N	all	none	none
L1	Y	Y	N	none	omit [MTA ∧ (ACDS)]	AH1
L2	Y	N	N	none	omit [MTA ∧ (ACDS)] lon always false	AH1
L3	Y	Y	Y	none	include [MTA ∧ (ACDS)]	AH1 and T1-T8 or TE1-TE8
L4	Y	N	Y	none	include [MTA ∧ (ACDS)] lon always false	AH1 and T1-T8 or TE1-TE8

C8. L Function (with Address Extension) Allowable Subsets

Identification	Description: Capabilities — Basic Extended Listener	Listen Only Mode	Unaddress If MSA ∧ (TPAS)*	States Omitted	Other Requirements	Other Function Subsets Required
LEØ	N	N	N	all	none	none
LE1	Y	Y	N	none	omit [MSA ∧ (TPAS) ∧ (ACDS)]	AH1
LE2	Y	N	N	none	omit [MSA ∧ (TPAS) ∧ (ACDS)] lon always false	AH1
LE3	Y	Y	Y	none	include [MSA ∧ (TPAS)] ∧ (ACDS)]	AH1 and T1-T8 or TE1-TE8
LE4	Y	N	Y	none	include [MSA ∧ (TPAS) ∧ (ACDS)] lon always false	AH1 and T1-T8 or TE1-TE8

*Replaced by MTA when used together with the T function.

C9. SR Function Allowable Subsets

Identification	Description	States Omitted	Other Requirements	Other Function Subsets Required
SRØ	no capability	all	none	none
SR1	complete capability	none	none	T1, T2, T5, T6, TE1, TE2, TE5 or TE6

C10. RL Function Allowable Subsets

Identification	Description	States Omitted	Other Requirements	Other Function Subsets Required
RLØ	no capability	all	none	none
RL1	complete capability	none	none	L1-L4 or LE1-LE4
RL2	no local lock out	LWLS and RWLS	rtl always false	L1-L4 or LE1-LE4

C11. PP Function Allowable Subsets

Identi-fication	Description	States Omitted	Other Requirements	Other Function Subsets Required
PPØ	no capability	all	none	none
PP1	remote configuration	none	include [((PPD ∧ (PACS) ∨ PPU) ∧ (ACDS)] include [PPE ∧ (PACS) ∧ (ACDS)] exclude lpe	L1-L4 or LE1-LE4
PP2	local configuration	PUCS, PACS	include lpe exclude [((PPD ∧ (PACS)) ∨ PPU) ∧ (ACDS)] exclude [PPE ∧ (PACS) ∧ (ACDS)] local messages shall be substituted for S, P1, P2, P3	none

C12. DC Function Allowable Subsets

Identi-fication	Description	States Omitted	Other Requirements	Other Function Subsets Required
DCØ	no capability	all	none	none
DC1	complete capability	none	none	L1-L4 or LE1-LE4
DC2	omit selective device clear	none	omit [SDC ∧ (LADS)]	AH1

C13. DT Function Allowable Subsets

Identi-fication	Description	States Omitted	Other Requirements	Other Function Subsets Required
DTØ	no capability	all	none	none
DT1	complete capability	none	none	L1-L4 or LE1-LE4

C14. C Function Allowable Subsets

Identi-fication*	Capabilities										Notes	States Required									Other Requirements			Other Function Subsets Required				
	System Controller	Send IFC and Take Charge	Send REN	Respond to SRQ	Send I.F. Messages	Receive Control	Pass Control	Pass Control to Self	Parallel Poll	Take Control Synchronously		SNAS, SACS	SIIS, SIAS, SINS	SRIS, SRAS, SRNS	CSNS, CSRS	CACS, CSBS, CSHS, CSWS, CAWS	CADS	CIDS	CTRS	CPWS, CPPS	[TCT ∧ $\overline{\text{ACDS}}$ ∧ $\overline{\text{TADS}}$]†	[$\overline{\text{TADS}}$]‡	*tcs* not always false	C1	C2	AH1, L1-L4, or LE1-LE4	SH1	T1-T8, TE1-TE8
C0	N	N	N	N	N	N	N	N	N	N		O	O	O	O	O	O	O	O	O	O	O	O	O	—	—	—	—
C1	Y	—	—	—	—	—	—	—	—	—	(1)	R	—	—	—	—	—	—	—	—	—	—	—	—	—	—	—	—
C2	—	Y	—	—	—	—	—	—	—	—	(1),(6)	—	R	—	—	—	—	—	—	—	—	—	—	R	—	—	—	—
C3	—	—	Y	—	—	—	—	—	—	—	(1)	—	—	R	—	—	—	—	—	—	—	—	—	R	—	—	—	—
C4	—	—	—	Y	—	—	—	—	—	—	(1)	—	—	—	R	—	—	—	—	—	—	—	—	—	—	—	—	—
C5	—	—	—	—	Y	Y	Y	Y	Y	Y	(2),(3)	—	—	—	—	R	R	R	R	R	R	R	R	—	—	R	R	R
C6	—	—	—	—	Y	Y	Y	Y	Y	N	(2),(3)	—	—	—	—	R	R	R	R	R	R	R	O	—	—	—	R	R
C7	—	—	—	—	Y	Y	Y	Y	N	Y	(2),(3)	—	—	—	—	R	R	R	R	O	R	R	R	—	—	R	R	R
C8	—	—	—	—	Y	Y	Y	Y	N	N	(2),(3)	—	—	—	—	R	R	R	R	O	R	R	O	—	—	—	R	R
C9	—	—	—	—	Y	Y	Y	N	Y	Y	(2),(3)	—	—	—	—	R	R	R	R	R	R	O	R	—	—	R	R	R
C10	—	—	—	—	Y	Y	Y	N	Y	N	(2),(3)	—	—	—	—	R	R	R	R	R	R	O	O	—	—	—	R	R
C11	—	—	—	—	Y	Y	Y	N	N	Y	(2),(3)	—	—	—	—	R	R	R	R	O	R	O	R	—	—	R	R	R
C12	—	—	—	—	Y	Y	Y	N	N	N	(2),(3)	—	—	—	—	R	R	R	R	O	R	O	O	—	—	—	R	R
C13	—	—	—	—	Y	Y	N	N	Y	Y	(2)	—	—	—	—	R	R	R	O	R	O	O	R	—	R	R	R	—
C14	—	—	—	—	Y	Y	N	N	Y	N	(2)	—	—	—	—	R	R	R	O	R	O	O	O	—	R	—	R	—
C15	—	—	—	—	Y	Y	N	N	N	Y	(2)	—	—	—	—	R	R	R	O	O	O	O	R	—	R	R	R	—
C16	—	—	—	—	Y	Y	N	N	N	N	(2)	—	—	—	—	R	R	R	O	O	O	O	O	—	R	—	R	—
C17	—	—	—	—	Y	N	Y	Y	Y	Y	(2),(3),(4)	—	—	—	—	R	O	R	R	R	R	R	R	—	—	R	R	R
C18	—	—	—	—	Y	N	Y	Y	Y	N	(2),(3),(4)	—	—	—	—	R	O	R	R	R	R	R	O	—	—	—	R	R
C19	—	—	—	—	Y	N	Y	Y	N	Y	(2),(3),(4)	—	—	—	—	R	O	R	R	O	R	R	R	—	—	R	R	R
C20	—	—	—	—	Y	N	Y	Y	N	N	(2),(3),(4)	—	—	—	—	R	O	R	R	O	R	R	O	—	—	—	R	R
C21	—	—	—	—	Y	N	Y	N	Y	Y	(2),(3),(4)	—	—	—	—	R	O	R	R	R	R	O	R	—	—	R	R	R
C22	—	—	—	—	Y	N	Y	N	Y	N	(2),(3),(4)	—	—	—	—	R	O	R	R	R	R	O	O	—	—	—	R	R
C23	—	—	—	—	Y	N	Y	N	N	Y	(2),(3),(4)	—	—	—	—	R	O	R	R	O	R	O	R	—	—	R	R	R
C24	—	—	—	—	Y	N	Y	N	N	N	(2),(3),(4)	—	—	—	—	R	O	R	R	O	R	O	O	—	—	—	R	R
C25	—	—	—	—	Y	N	N	N	Y	Y	(2),(5)	—	—	—	—	R	O	O	O	R	O	O	R	—	—	R	R	—
C26	—	—	—	—	Y	N	N	N	Y	N	(2),(5)	—	—	—	—	R	O	O	O	R	O	O	O	—	—	—	R	—
C27	—	—	—	—	Y	N	N	N	N	Y	(2),(5)	—	—	—	—	R	O	O	O	O	O	O	R	—	—	R	R	—
C28	—	—	—	—	Y	N	N	N	N	N	(2),(5)	—	—	—	—	R	O	O	O	O	O	O	O	—	—	—	R	—

*Typical notation to describe a controller consists of the letter C followed by one or more of the numbers indicating the subsets selected. For example; C1, 2, 3, 4, 8.

†This is part of the CIDS to CADS transitional expression.

‡This is part of the CACS to CTRS transitional expression.

NOTES:
(1) One or more of subsets C1 through C4 may be chosen in any combination with any one of C5 through C28.
(2) Only one subset may be chosen from C5 through C28.
(3) The CTRS state must be included in devices which are to be operated in multicontroller systems.
(4) These subsets are not allowed unless C2 is included.
(5) These subsets are intended to be used in devices and systems where no control passage is possible.
(6) When a system controller asserts IFC during the time another physical device is operating as controller-in-charge, the system controller should refrain from active assertion of the source handshake and ATN until the removal of the IFC message to preclude multiple controller contention.
O = omit, R = required, hyphen = not applicable or not required, Y = yes, N = no.

Appendix D
Interface Message Reference List

Mnemonic	Message	Interface Function(s)
LOCAL MESSAGES RECEIVED (by interface functions)		
gts	go to standby	C
ist	individual status qualifier	PP
lon	listen only	L, LE
[lpe]	local poll enable	PP
ltn	listen	L, LE
lun	local unlisten	L, LE
nba	new byte available	SH
pon	power on	SH, AH, T, TE, L, LE, SR, RL, PP, C
rdy	ready	AH
rpp	request parallel poll	C
rsc	request system control	C
rsv	request service	SR
rtl	return to local	RL
sic	send interface clear	C
sre	send remote enable	C
tca	take control asynchronously	C
tcs	take control synchronously	AH, C
ton	talk only	T, TE
LOCAL MESSAGES SENT (to device functions)		
	None defined; see Message Output tables in Section 2 for description of Device Function Interaction which provides guidelines as to the appropriate states from which local messages may be sent to the device functions.	
REMOTE MESSAGES RECEIVED		
ATN	attention	SH, AH, T, TE, L, LE, PP, C
DAB	data byte	(via L, LE)
DAC	data accepted	SH
DAV	data valid	AH
DCL	device clear	DC
END	end	(via L, LE)
GET	group execute trigger	DT
GTL	go to local	RL
IDY	identify	L, LE, PP
IFC	interface clear	T, TE, L, LE, C
LLO	local lockout	RL
MLA	my listen address	L, LE, RL
[MLA]	my listen address	T
MSA or [MSA]	my secondary address	TE, LE
MTA	my talk address	T, TE
[MTA]	my talk address	L
OSA	other secondary address	TE
OTA	other talk address	T, TE
PCG	primary command group	TE, LE, PP
PPC	parallel poll configure	PP
[PPD]	parallel poll disable	PP
[PPE]	parallel poll enable	PP
PPRn	parallel poll response n	(via C)
PPU	parallel poll unconfigure	PP
REN	remote enable	RL
RFD	ready for data	SH
RQS	request service	(via L, LE)

Interface Message Reference List
(Continued)

Mnemonic	Message	Interface Function(s)
[SDC]	selected device clear	DC
SPD	serial poll disable	T, TE
SPE	serial poll enable	T, TE
SRQ	service request	(via C)
STB	status byte	(via L, LE)
TCT or [TCT]	take control	C
UNL	unlisten	L, LE
REMOTE MESSAGES SENT		
ATN	attention	C
DAB	data byte	(via T, TE)
DAC	data accepted	AH
DAV	data valid	SH
DCL	device clear	(via C)
END	end	(via T)
GET	group execute trigger	(via C)
GTL	go to local	(via C)
IDY	identify	C
IFC	interface clear	C
LLO	local lockout	(via C)
MLA or [MLA]	my listen address	(via C)
MSA or [MSA]	my secondary address	(via C)
MTA or [MTA]	my talk address	(via C)
OSA	other secondary address	(via C)
OTA	other talk address	(via C)
PCG	primary command group	(via C)
PPC	parallel poll configure	(via C)
[PPD]	parallel poll disable	(via C)
[PPE]	parallel poll enable	(via C)
PPRn	parallel poll response n	PP
PPU	parallel poll unconfigure	(via C)
REN	remote enable	C
RFD	ready for data	AH
RQS	request service	T, TE
[SDC]	selected device clear	(via C)
SPD	serial poll disable	(via C)
SPE	serial poll enable	(via C)
SRQ	service request	SR
STB	status byte	(via T, TE)
TCT	take control	(via C)
UNL	unlisten	(via C)
UNT	untalk	(via C)

Appendix E
Multiline Interface Messages: ISO Code Representation

(SENT AND RECEIVED WITH ATN=1)

b_4	b_3	b_2	b_1	Bits b_7 b_6 b_5 → / ROW ↓ COLUMN →	0 0 0 (0)	MSG (1)	0 0 1 (1)	MSG	0 1 0 (2)	MSG	0 1 1 (3)	MSG	1 0 0 (4)	MSG	1 0 1 (5)	MSG	1 1 0 (6)	MSG	1 1 1 (7)	MSG
0	0	0	0	0	NUL		DLE		SP	MLA ASSIGNED TO DEVICE	0	MLA ASSIGNED TO DEVICE	@	MTA ASSIGNED TO DEVICE	P	MTA ASSIGNED TO DEVICE		MEANING DEFINED BY PCG CODE	p	MEANING DEFINED BY PCG CODE
0	0	0	1	1	SOH	GTL	DC1	LLO	!		1		A		Q		a		q	
0	0	1	0	2	STX		DC2		"		2		B		R		b		r	
0	0	1	1	3	ETX		DC3		#		3		C		S		c		s	
0	1	0	0	4	EOT	SDC	DC4	DCL	$		4		D		T		d		t	
0	1	0	1	5	ENQ	PPC (3)	NAK	PPU	%		5		E		U		e		u	
0	1	1	0	6	ACK		SYN		&		6		F		V		f		v	
0	1	1	1	7	BEL		ETB				7		G		W		g		w	
1	0	0	0	8	BS	GET	CAN	SPE	(		8		H		X		h		x	
1	0	0	1	9	HT	TCT	EM	SPD	)		9		I		Y		i		y	
1	0	1	0	10	LF		SUB		*		:		J		Z		j		z	
1	0	1	1	11	VT		ESC		+		;		K		[		k		{	
1	1	0	0	12	FF		FS		,		<		L		\		l		¦	
1	1	0	1	13	CR		GS		–		=		M		]		m		}	
1	1	1	0	14	SO		RS		.		>		N		^		n		~	
1	1	1	1	15	SI		US		/		?	UNL	O		___	UNT	o		DEL	

ADDRESSED COMMAND GROUP (ACG) — columns 0 and MSG

UNIVERSAL COMMAND GROUP (UCG) — column 1 and MSG

LISTEN ADDRESS GROUP (LAG) — columns 2 and 3

TALK ADDRESS GROUP (TAG) — columns 4 and 5

PRIMARY COMMAND GROUP (PCG) — columns 0 through 5

SECONDARY COMMAND GROUP (SCG) — columns 6 and 7

(4) — columns 2 through 5

NOTES:
(1) MSG = INTERFACE MESSAGE
(2) b_1 = DIO1 . . . b_7 = DIO7
(3) REQUIRES SECONDARY COMMAND
(4) DENSE SUBSET (COLUMN 2 THROUGH 5)

Fig E1
Multiline Interface Messages: ISO-7 Bit Code Representation

Appendix F
Logic Circuit Implementation

To assist the designer in the interpretation of the state diagrams, possible circuit implementations are given for situations occurring within the interface functions. It must be understood that the logic diagrams given in this appendix do not show the only implementations possible nor do they even represent recommended implementations. They are for educational purposes only.

State diagrams are used to represent two concepts:

(1) They allow differentiation between the different responses an interface function might produce and identify each with one or more unique states of the interface function.

(2) They identify those situations where an interface function is required to remember past events in order to produce the correct response.

Each state in any of the diagrams serves either or both of these purposes. For example, the LADS of the L interface function has no unique response associated with it and cannot be distinguished from LIDS. Its purpose, however, is to remember that the device has received a listen address over the bus and is therefore able to enter LACS when the ATN message is received false (see Fig 7). Conversely, the LACS is an example of one which has no memory with it but which exists as a distinct state only to show a specific response capability which does not exist during LADS. The only internal difference between the two states is the value of the ATN message, and no memory is required since this message value is continuously available.

F1. Implementing States Which Require No Memory

The DT interface function is an example of a complete interface function which requires no memory. Its state diagram (identical to Fig 13) is the following:

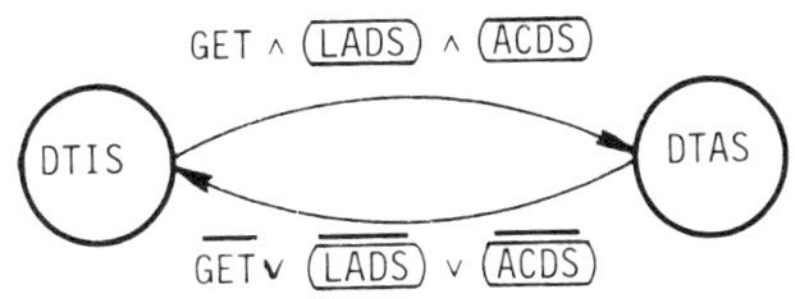

Since no memory is required, this interface function can be implemented with a single AND gate:

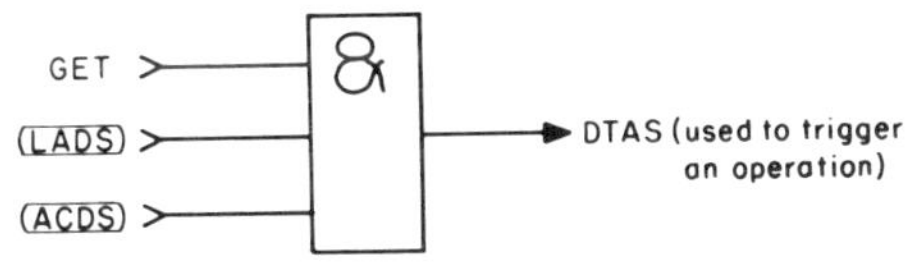

F2. Implementing States Which Require Memory

The SR interface function is an example of one requiring memory for its states. Its state diagram (identical to Fig 9) is the following:

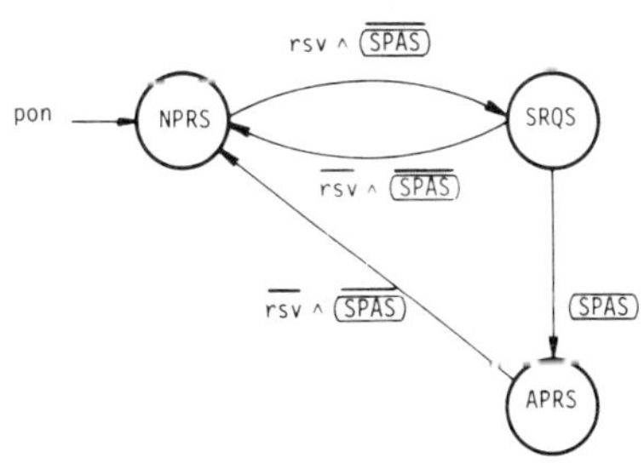

The two top states taken by themselves represent a circuit whose internal state follows

(a) (b)

Fig F1
Partial SR Function —
(a) Composite Logic; (b) Gate Level Logic

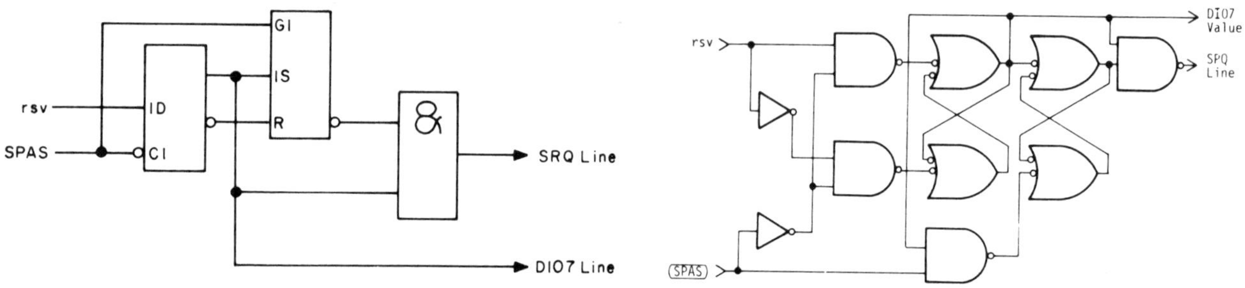

(a) (b)

Fig F2
Full SR Function —
(a) Composite Logic; (b) Gate Level Logic

the value of the rsv message but only if SPAS is false. Fig F1 is a standard DC Flip-Flop.

To complete the circuit all that is needed is a memory that (SPAS) has occured after the latch has turned on. This circuit can be built around a standard RS flip-flop and added to the latch to produce Fig F2.

In this circuit, the RS flip-flop output stage is forced clear whenever the value of the latched rsv message is false. When the latched rsv message becomes true, it remains cleared until the first time SPAS becomes active at which time it sets, remembering that an RQS message has been sent and SRQ no longer needs to be held true.

Appendix G
Parallel Polling Sequence

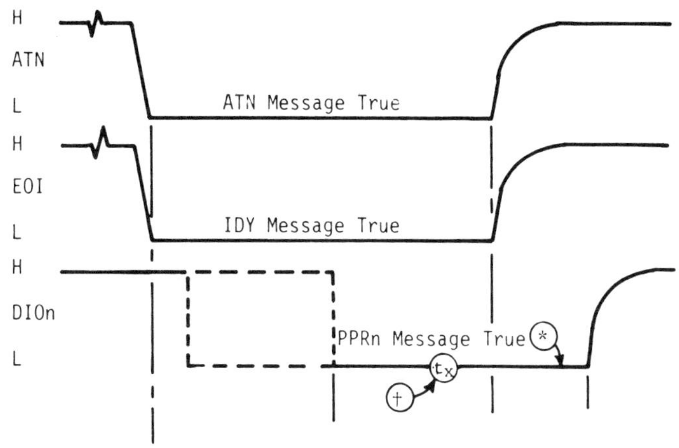

Fig G1
Parallel Poll Sequence: Signal Line Waveforms

*PPRn message true shown in one of two alternate states as determined by the PPE message.

†Strobe of DIOn lines occurs internal to the controller at any time during the CPPS state by a designer defined method (the transfer of status data bits during parallel poll does not utilize the handshake process).

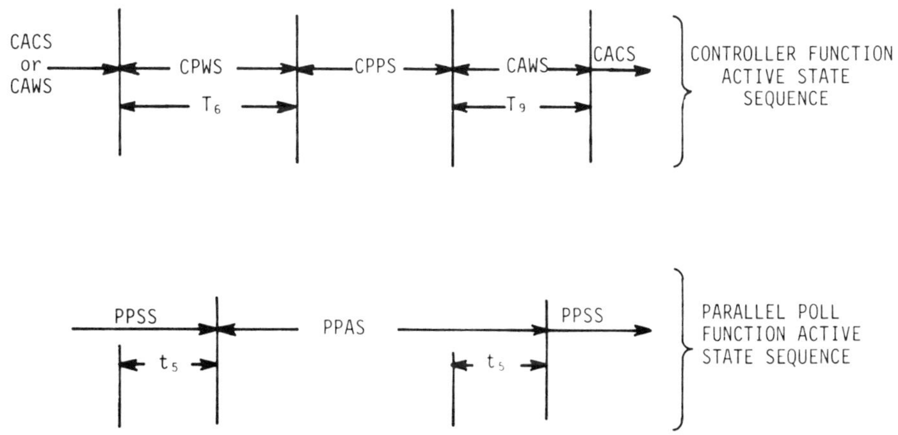

Fig G2
Interface Function Active States During Parallel Poll

Page 42 Table 34. *The following item has been added to the right-hand column:* CSHS = controller standby hold state

2.12.2 C Function State Diagram
Page 43 Fig 14. *The following item has been added to the state diagram:* **CSHS.** See Page 3, revised Fig 14 (relevant transition terms also revised).

Page 44 Table 35. *The following item has been added to the table:*

C State	ATN	IDY	Multiline	Device Function DF Interaction
CSHS	F	(F)	(NUL)	DF shall not send interface messages

2.12.3.6 CSBS (Controller Standby State)
Page 45. *The last paragraph of* **2.12.3.6** *has been changed to read as follows:* The C function shall exit CSBS and enter:

(1) The CSHS (controller standby hold state) if the tcs (take control synchronously) message is true and ANRS is active

(2) The CSWS (controller synchronous wait state) if the tca (take control asynchronously) message is true

(3) The CIDS within t_4 if the IFC message is true and SACS is not active

2.12.3.7 CSWS (Controller Synchronous Wait State)
Page 46. *The last paragraph of the exit condition for CSWS has been changed to read as follows:*

The C function shall exit CSWS and enter:

(1) The CAWS after a period of T_7 has elapsed or if TADS is active

2.12.3.20 CSHS (Controller Standby Hold State)
Page 47. *A new clause has been added to C function description. It reads as follows:*

In CSHS the C function is in a hold state until such time as the DAV message is false as presented to all devices connected to the system. The CSHS state prevents false coincidence of the ATN and DAV messages, as observed by idle devices, during a tcs sequence.

In CSBS the ATN message shall be sent false, the IDY message shall be sent passive false, and the NUL message shall be sent passive true.

The C function shall exit CSHS and enter:

(1) The CSWS (Controller Standby Wait State) if a period of T_{10} has elapsed.

(2) The CSBS if the tcs message is false,

(3) The CIDS within t_4 if the IFC message is true and SACS is not active.

2.12.4 C Interface Function Allowable Subsets
Page 48 Table 37. CSHS *has been added to* **Table 37** *under* **States Required,** *5th column. It reads as follows:*

CACS, CSBS, CSHS, CSWS, CAWS

3.8 State Transition Timing Values
Page 55 Table 39. T_{10} *has been added to* **Table 39.** *It reads as follows:*

Time Value	Function	Description	Value
T_{10}	C	delay for $\overline{\text{DAV}}$	$\geq 1.5\ \mu s$

Appendix C
Page 77 Table C14. CSHS *has been added to* **Table C14** *under* **States, Required,** *5th column. It reads as follows:*

CACS, CSBS, CSHS, CSWS, CAWS

NOTE: The text changes associated with supplement IEEE Std 488A-1980 are identified by a double vertical line in the margin opposite the relevant subject matter in this printing.

An American National Standard

IEEE Recommended Practice for Code and Format Conventions

For Use with ANSI/IEEE Std 488-1978, IEEE Standard Digital Interface for Programmable Instrumentation

Sponsor

IEEE Instrumentation and Measurements Society

Approved December 18, 1981

IEEE Standards Board

Approved April 15, 1983

American National Standards Institute

IEEE Standards documents are developed within the Technical Committees of the IEEE Societies and the Standards Coordinating Committees of the IEEE Standards Board. Members of the committees serve voluntarily and without compensation. They are not necessarily members of the Institute. The standards developed within IEEE represent a consensus of the broad expertise on the subject within the Institute as well as those activities outside of IEEE which have expressed an interest in participating in the development of the standard.

Use of an IEEE Standard is wholly voluntary. The existence of an IEEE Standard does not imply that there are no other ways to produce, test, measure, purchase, market, or provide other goods and services related to the scope of the IEEE Standard. Furthermore, the viewpoint expressed at the time a standard is approved and issued is subject to change brought about through developments in the state of the art and comments received from users of the standard. Every IEEE Standard is subjected to review at least once every five years for revision or reaffirmation. When a document is more than five years old, and has not been reaffirmed, it is reasonable to conclude that its contents, although still of some value, do not wholly reflect the present state of the art. Users are cautioned to check to determine that they have the latest edition of any IEEE Standard.

Comments for revision of IEEE Standards are welcome from any interested party, regardless of membership affiliation with IEEE. Suggestions for changes in documents should be in the form of a proposed change of text, together with appropriate supporting comments.

Interpretations: Occasionally questions may arise regarding the meaning of portions of standards as they relate to specific applications. When the need for interpretations is brought to the attention of IEEE, the Institute will initiate action to prepare appropriate responses. Since IEEE Standards represent a consensus of all concerned interests, it is important to ensure that any interpretation has also received the concurrence of a balance of interests. For this reason IEEE and the members of its technical committees are not able to provide an instant response to interpretation requests except in those cases where the matter has previously received formal consideration.

Comments on standards and requests for interpretations should be addressed to:

Secretary, IEEE Standards Board
345 East 47th Street
New York, NY 10017
USA

Foreword

(This Foreword is not a part of IEEE Std 728-1982, IEEE Recommended Practice for Code and Format Conventions for Use with ANSI/IEEE Std 488-1978, IEEE Standard Digital Interface for Programmable Instrumentation.)

This recommended practice is written as a companion for and extension to ANSI/IEEE Std 488-1978. The reader should be thoroughly familiar with ANSI/IEEE Std 488 to best understand and utilize the content of this recommended practice. Though written from the perspective of the device designer, users who design, assemble, or program systems will find this recommended practice helpful.

Since this recommended practice elaborates several different alternatives to accomplish equivalent or similar message transfers it is identified as a recommended practice (guidelines) rather than a standard (in absolute terms). Information interchange applicable to ANSI/IEEE Std 488-1978 has not yet matured to the point where a meaningful standard on codes and formats can be written. This recommended practice defines what is considered to be a useful set of code and format conventions. Users may inform the IEEE Standards office of any significant problems encountered in the use of this recommended practice by writing to:

Secretary
IEEE Standard Board
345 East 47th Street
New York, N.Y. 10017

Relation to IEC Publication 625-2

This IEEE recommended practice deals, in part, with the same subject matter as IEC Publication 625-2, however, the content of these documents is not identical. While the technical content of both documents are in general agreement where the subject matter is similar, the use of syntax diagrams and definitions contained in this recommended practice is thought to add significant clarity. In addition, this recommended practice contains recommendations for the format of character, string and block data not now contained in IEC 625-2. Users interested in interconnecting products designed to these two documents are cautioned to examine both the documents and related products in detail.

ANSI/IEEE Std 488-1978, IEEE Standard Digital Interface for Programmable Instrumentation sets forth the device-independent mechanical, electrical and functional specifications necessary to support the configuration of programmable instruments using a byte-serial bit-parallel means to transfer digital data among interconnected instruments and apparatus. A fourth category of specifications dealing with the more operational aspects of an interface system is necessary to enable the configuration of an operational system. Operational specifications for systems and devices tend to be more device and system dependent. Program and measurement message formats, application software, and diagnostic routines, are examples of operational related capabilities. This recommended practice addresses the structure and format of several message types needed to support the more operational aspects of systems. No single code set and message format appears to be feasible, for use with ANSI/IEEE Std 488-1978 and still support the overall objective of permitting a wide range of products and product capabilities to be interfaced. It is feasible, however, to define a limited set of guidelines and alternatives setting forth a number of different codes and formats generally applicable to products implemented with ANSI/IEEE Std 488-1978 capability. The "Code and Format Conventions" of this recommended practice reflect the general customs and practice found useful for the exchange of device-dependent messages.

The word *code* as used herein refers to the pattern of bits in a device-dependent data byte (DAB message). Unless otherwise stated, the 7-bit coded character set used in this recommended practice is the ASCII 7-bit code.* The ASCII 7-bit code and its relation to the DIO lines of ANSI/IEEE Std 488-1978 is defined in Sections 10 and 11.

The word *format* as used herein refers to the organization of data bytes within commonly used device-dependent messages.

---

*ANSI X3.4-1977, American National Standard Code for Information Interchange. The ISO 7-bit code is the equivalent international reference version.

This recommended practice uses the same letter mnemonics (for example, END, SPAS, DAB) to refer to remote messages and interface states as used throughout ANSI/IEEE Std 488-1978. Additionally, this recommended practice refers to several non-printing ASCII characters that are represented in text form by two or three characters although the message itself is both represented and sent by a single multiline coded character (for example, new line represented by NL). These coded characters are identified throughout this recommended practice by underlined text (for example, CR, LF, NL, ETX).

The two IEEE Committees responsible for the preparation of this standard within the US were the Instrumentation and Measurements Society Subcommittee on Instrument/Computer Interfaces (which also served as the US Advisory Committee to US representatives on IEC TC 66/WG3 but is now SC 65C/WG3) and the Instrumentation and Measurements Society Technical Committee on Automated Instrumentation.

Personnel for the Instrument/Computer Interfaces Subcommittee were:

John Peatman, *Chairman* **D. C. Loughry, *Secretary***

R. Cram
R. Fulks
R. Gould
T. Leedy
W. Maciejewski
Z. Tarczy-Hornoch
J. Pepin
J. Skeen
L. Sollman
D. Trout
D. Ware

When the IEEE Standards Board approved this standard on December 18, 1981, it had the following membership:

I. N. Howell, Jr, *Chairman* **Irving Kolodny, *Vice Chairman***

Sava I. Sherr, *Secretary*

G. Y. R. Allen
J. J. Archambault
J. H. Beall
J. T. Boettger
Edward Chelotti
Edward J. Cohen
Len S. Corey
Jay Forster
Kurt Greene
Loering M. Johnson
Joseph L. Koepfinger
J. E. May
Donald T. Michael*
J. P. Riganati
F. Rosa
R. W. Seelbach
J. S. Stewart
W. E. Vannah
Virginius N. Vaughan, Jr
Art Wall
Robert E. Weiler

*Member emeritus

Contents

An American National Standard

IEEE Recommended Practice for Code and Format Conventions

For Use with ANSI/IEEE Std 488-1978,
IEEE Standard Digital Interface for
Programmable Instrumentation

1. Scope and Object

1.1 Scope. This recommended practice elaborates a family of codes and formats to be generated, processed, and interpreted by the device functions of apparatus operating in concert with interface functions.

This recommended practice is intended to apply to both those devices with limited ability to generate, process, or interpret a variety of codes and formats as well as those devices that have extensive ability to generate, process, and interpret unique and very specialized codes and formats.

This recommended practice applies to the organization and content of messages at the machine-to-machine information interchange level, however, the guidelines set forth herein may have critical impact on both the related person-to-machine and machine-to-person information interchange.

This recommended practice applies to several device-dependent message types. Figure 1 shows the major relationships between this recommended practice and the related parts of ANSI/IEEE Std 488-1978 [3].[1]

This recommended practice is intended to apply directly to instrumentation systems and may also apply to certain devices outside the scope of the instrument system environment.

[1]The number in brackets corresponds to the references in Section 2 of this recommended practice.

1.2 Object. The objectives of this recommended practice are:

(1) To promote compatibility among different manufacturers' products

(2) To enable the interconnection of instrumentation and related devices with both limited and extensive capability to generate, process, and interpret a variety of different message types

(3) To define codes and formats that will minimize the generation of application software and system configuration costs

(4) To define a limited family of preferred message codes and formats in a relatively device-independent manner

(5) To permit direct communication among instrument system devices without extraordinary translation and conversion of special codes and formats

2. References

[1] ANSI X3.4-1977, American National Standard Code for Information Interchange.[2]

[2] ANSI X 3.42-1975, American National Standard Representation of Numeric Values in Character Strings for Information Interchange.

[3] ANSI/IEEE Std 488-1978, IEEE Standard Digital Interface for Programmable Instrumentation.

[2]ANSI documents are available from the Sales department of American National Standards Institute, 1430 Broadway, New York, NY 10018.

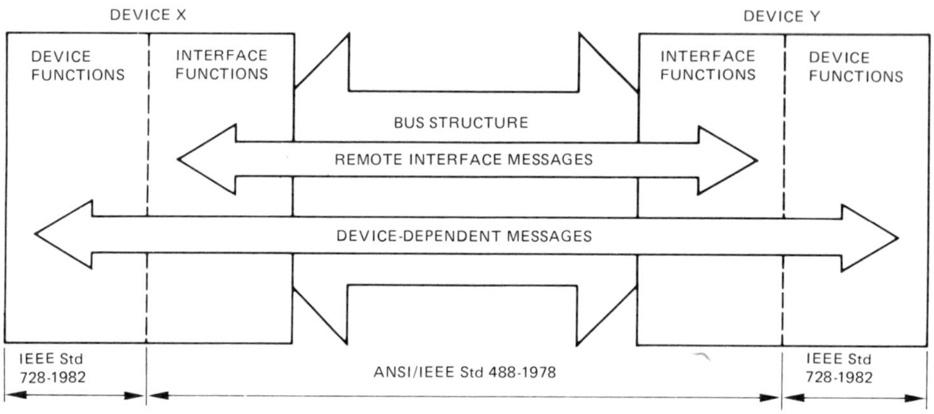

Fig 1
Functions and Messages of the Interface System

3. System Considerations

The purpose of an interface system is to carry device-dependent messages between devices. According to interface functions, to and from device functions as shown in Fig 1. Device functions may differ from one another in several respects, and thus affect the nature of device-dependent messages that may be interchanged. The type and complexity of messages interchanged is, in part, determined by these factors:

(1) Ability of devices to generate, process, or interpret only a limited or single set of codes and formats versus an extensive or complex set of data codes and formats

(2) Data transferred at machine level versus data initiated or consumed according to human contact

The type of system or systems a given product is to be placed in (that is, other products it must communicate with and under what circumstances) has a significant impact on the nature of device-dependent messages that the product may be capable of generating, processing, or interpreting. The codes and formats used by a given product will determine, to a large extent, the range of configurations over which that product can be easily integrated. Therefore, careful consideration should be given to a device's code and format capability to ensure satisfactory system performance.

A designer should consider the range of products a given device is expected to communicate with and view potential applications from a broad perspective. Accommodation of more than one code or format may be necessary. Compatability with a variety of products may be increased if messages are received *forgivingly* or interpreted without strict compliance to a specific format (for example, accept variable data field lengths, ignore leading spaces). See Appendix D.

4. Message Concepts

Most of the device-dependent messages described by this recommended practice are composed of a sequence of data bytes (DAB). A common structure and organization of the data byte sequence promotes compatibility among different products. This recommended practice sets forth the general organization and format for device-dependent messages and leaves the specific content of such messages to the designer.

4.1 Message Types and Definitions. Devices typically send and receive several different types of device-dependent messages. This section defines a set of message types considered to be useful for a variety of instrumentation system needs. It also defines the terms used to structure the description of these messages.

4.1.1 message. A message, as used throughout this recommended practice, refers to one of four basic types: measurement, program, status and display. The messages described herein are additional to and in concert with the terms *interface message, remote message, local message,* and *device-dependent message* as used throughout ANSI/IEEE Std 488-1978. [3]

4.1.1.1 measurement message. A measurement message is a sequence of one or more measurement results separated by message unit separators. See Section 5.

4.1.1.2 program message. A program message is a sequence of one or more program instructions and optional message unit separators. See Section 6.

4.1.1.3 status byte message. A status byte message carries information about the internal condition(s) of an instrument's device functions. Devices send a status byte (STB message) in response to a serial poll to indicate one or more device function conditions (for example, measurement complete, out of calibration, current limited). A status byte message is sent with a single status byte. See Section 7.

4.1.1.4 display message. A display message contains information displayed for operator convenience. Display data may contain a series of accumulated measurement results or a series of program instructions used to setup a device. In either case, human interpretation is important. See Sections 5, 6, and 8.

4.1.2 message unit. A message unit is a single measurement result or a single program instruction.

4.1.2.1 measurement result. A measurement result is the output of an instrument's measurment process. Instrumentation devices typically output measurement information (for example, frequency, voltage, current) as a result of performing a measurement. The general format of measurement results may contain both alpha information and a numeric value. A measurement result consists of a sequence of DABs that represent an optional header field and one mandatory data field. See Section 5.

4.1.2.2 program instruction. A program instruction is used to setup and execute an instrument's measurement or stimulus function(s). Instrumentation devices typically receive program data (for example, measurement range, operating mode, output mode) in preparation for performing a measurement function. A program instruction consists of a sequence of DABs that represents at least a mandatory header field and one or more optional data fields. See Section 6.

4.1.3 message unit separator. A message unit separator is always used to separate multiple measurement results from one another. It is optionally used to separate multiple program instructions from one another.

4.1.4 message element. Message elements are the basic *building blocks* of the syntactic constructs that define program and measurement messages. Major message elements consist of three types of headers, four data types, and three types of separators. Specific message elements are defined in 4.3.

4.2 Message Overview. A device-dependent message of the type listed in 4.1 represents an amount of information whose beginning and end can be defined or implied and whose contents are generated, carried, and interpreted together as a unit. These messages are considered to be further partitioned into specific fields to reflect identifiable portions of the message within a message unit.

In general, a device-dependent message is composed of several message elements or fields such as a header field (typically alpha), data field (typically numeric), and separator field (typically a symbolic character). Not all messages need to contain all possible message elements. Each of the message types identified in 4.1.2.1 and 4.1.2.2 may, for example, contain a different set of message elements for different applications to which the instrument or device is applied. Further, messages may be either isolated or grouped to form a continuous flow of messages as the specific application demands.

4.2.1 Simplified Message Structure. Figure 2 presents a very simplified structure for two common message type measurement messages and program messages. The header field appears first in the sequence of DABs and is used to identify the nature of the value given in the data field of the message. The data field may contain numeric, string, character or block data types.

The separator field is considered as a separate entity and is used by the receiving device to aid in the processing of the overall message. The separator is considered as optional and application dependent for certain message constructs as defined in Sections 5 and 6.

For convenience in further descriptions, each major message element is identifiable by letter designations: Header Field (HR), Data Representation Field (DR), and Separator Field (SR). Subsequent descriptions will further expand this notation for use in identifying the several format

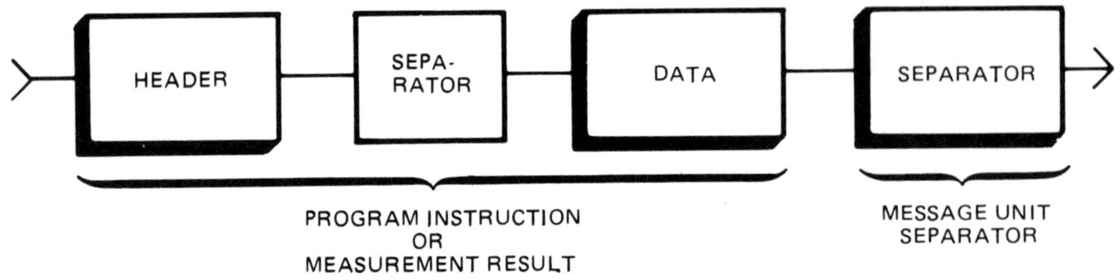

Fig 2
Simplified Message Structure

capabilities employed in various instruments. See Appendix B.

Status byte messages are sent within a single byte and hence do not utilize the same data fields. Instead, subfields all within the STB message specified by ANSI/IEEE Std 488-1978 [3] are defined throughout Section 7.

4.2.2 Syntax Diagram Notation. The generation and processing of measurement and program messages can be facilitated by defining the set of formats found most useful in instrumentation systems. Message format structure can be symbolized by syntactic diagrams that indicate allowable alternatives for the construction of specific messages from a sequence of data bytes. While the exact content of all data bytes is beyond the scope of this recommended practice, use of these general message formats should facilitate the interchange of information among devices.

The syntax diagram is a formal means to describe the preferred alternatives for constructing measurement or program messages. Each of the major message elements (HR, DR, SR) and their respective sets can be used to assemble messages for unambiguous processing and interpretation.

Devices, instruments in particular, are expected to send device-dependent data (in TACS) according to the message formats defined throughout this recommended practice. Both instruments and controllers can improve the level of information interchange by listening *forgivingly* when interpreting the message. See Appendix D for specific recommendations.

The syntax diagrams defined throughout Sections 4, 5, and 6 provide the basic definitions and message format structure for this recommended practice, further written description and examples serve to clarify and illustrate the use of these basic diagrams.

Syntax diagrams, comprised of six essential symbols, are used to define most of the message elements throughout this recommended practice:

(1) Rectangular shaped symbols indicate individual message elements that are explicitly defined elsewhere throughout this Section.

(2) Circular shaped symbols indicate literal data (for example, a specific ASCII character) the coding of which is defined in Section 10.

(3) Arrows, left to right, indicate the assembly sequence of successive message elements and the by-pass paths around the indicated message element(s).

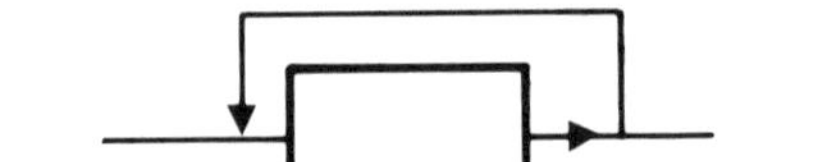

(4) Reverse arrows, right to left, indicate feedback loops for the repetition of one or more message elements.

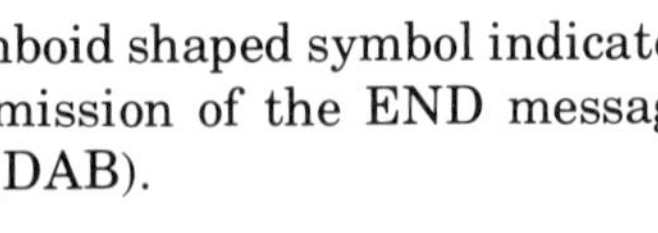

(5) Rhomboid shaped symbol indicates concurrent transmission of the END message with a data byte (DAB).

(6) Heavy lines indicate a preferred path within a syntax diagram, (see relevant text for explanation of preference).

In general, flow through a syntax diagram proceeds from left to right. Strategic location of arrowheads precludes disallowed paths through the diagram. Any circular or rectangular symbol without an exit path (line) implies the last byte of a message. Sections 4.3 through 4.3.3.5 define the sets of major message elements available. Sections 5 and 6 utilize these building blocks to establish the preferred sets of measurement and program messages respectively.

4.3 Message Unit Elements

4.3.1 Header Field. A header field may be used to describe the units(type, quantity) and quality of the data present in the data field it precedes, or both. A header field may also be used to select a specific function.

In all cases, the initial character in an HR field is limited to an alpha character to facilitate parsing (partitioning) of messages by the receiving device. Three header fields available as shown in Fig 3 are HR1 an alpha header, HR2 a formatted header, HR3 a character header.

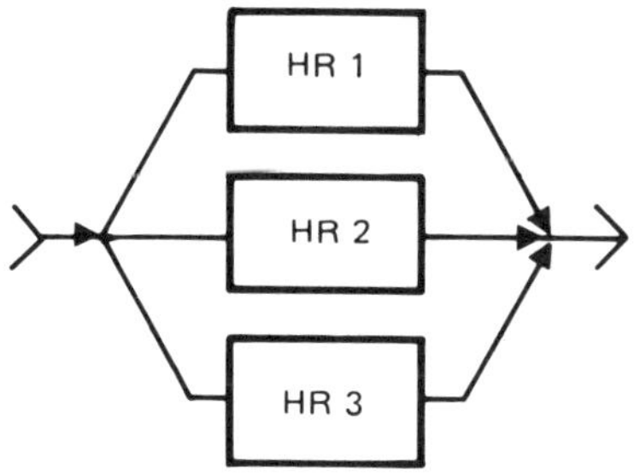

Fig 3
HR Choices

4.3.1.1 Alpha Header (HR1). A sequence of one or more alpha characters constructed according to Fig 4. Alpha characters may be of either case, upper case preferred.

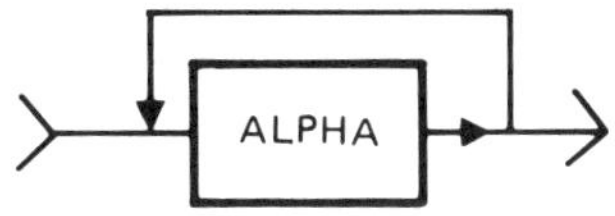

Fig 4
HR1 Syntax Diagram

| A | B | C | D | E | F | G | H | I | J | K | L | M |
N | O | P | Q | R | S | T | U | V | W | X | Y | Z |

| a | b | c | d | e | f | g | h | i | j | k | l | m | n | o |
p | q | r | s | t | u | v | w | x | y | z |

NOTE: The symbol | | is used here and throughout this recommended practice, to delineate the specified codes and, of itself, is not part of the code set.

4.3.1.2 Formatted Header (HR2). A sequence of one or more alpha characters plus the possibility of embedded or trailing spaces constructed according to Fig 5. In some messages it is helpful to maintain the header at a fixed length for a given product employing a variety of type and quality indications. The use of embedded or trailing spaces is therefore useful.

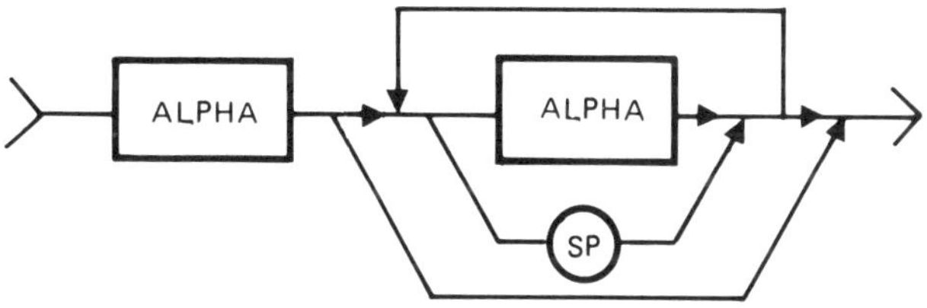

Fig 5
HR2 Syntax Diagram

NOTE: The symbol | SP | is used here, and throughout this recommended practice to indicate a *space* (ASCII code in column 2 row 0 of Table 5).

4.3.1.3 Character Header (HR3). A sequence of one alpha character plus the possibility of additional characters following an initial alpha constructed according to Fig 6. In the additional character set these printable characters are allowed:

Special characters:
| ! | $ | % | & | ' | (|) | * | + | - | . | / | : | < |
= | > | | ? | @ | [| \ |] | ^ | _ | \ | { | | | }
| ~ |

Alpha characters:
[Upper and lower case as listed in 4.3.1.1]

Digits:
| 0 | 1 | 2 | 3 | 4 | 5 | 6 | 7 | 8 | 9 |

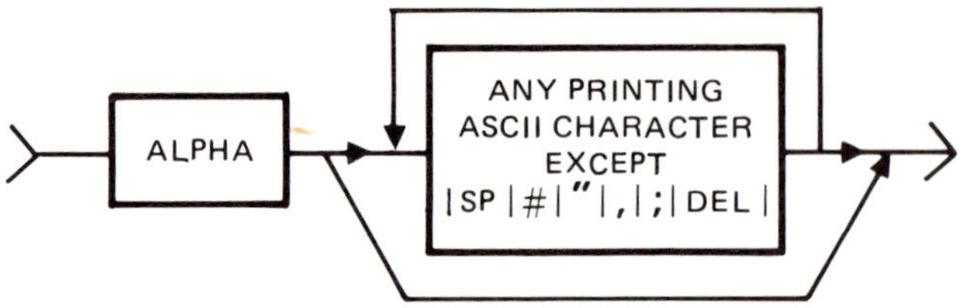

Fig 6
HR3 Syntax Diagram

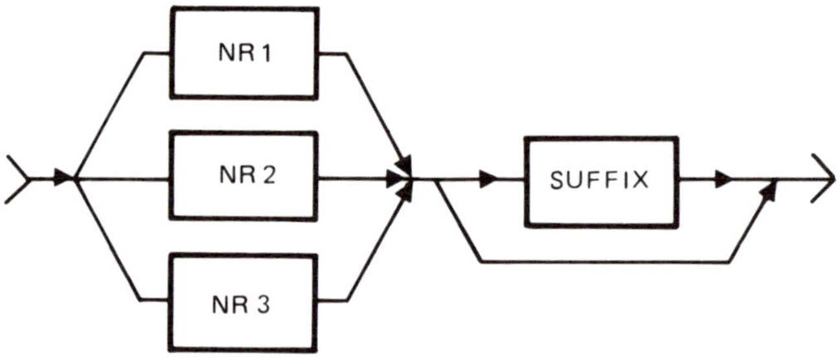

Fig 7
Numeric Representation Syntax Diagram

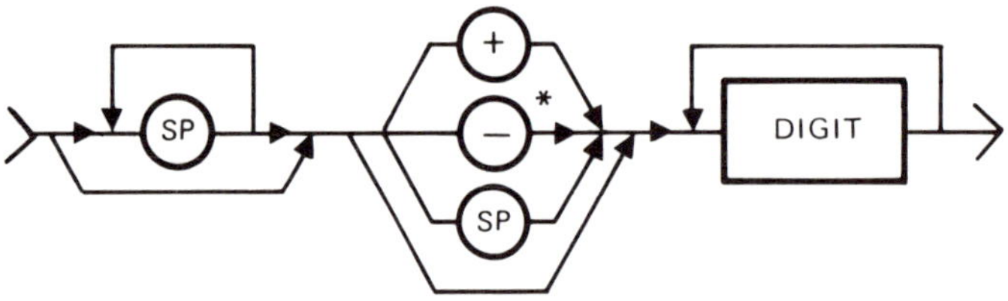

*Not preferred for ZERO valued numeric

Fig 8
NR1 Syntax Diagram

4.3.2 Data Fields. The central information contained within the body of each message unit may be represented by using one of four possible data types: numeric, string, block, or character. Some of these data types may be further subdivided into more specific categories. This subsection sets forth the detailed syntactic description for each data field type. Subsequent sections utilize these building blocks to construct measurement and program messages.

4.3.2.1 Numeric Data Type. The decimal positional representation of numeric values, commonly called numeric representation (NR), may be implemented in any of three forms as shown in Fig 7: NR1, NR2, NR3. These representations in general follow ANSI X3.42-1975 [2].

The NR1, NR2, NR3 representation applies only to the radix 10 with syntactic rules for the character set:

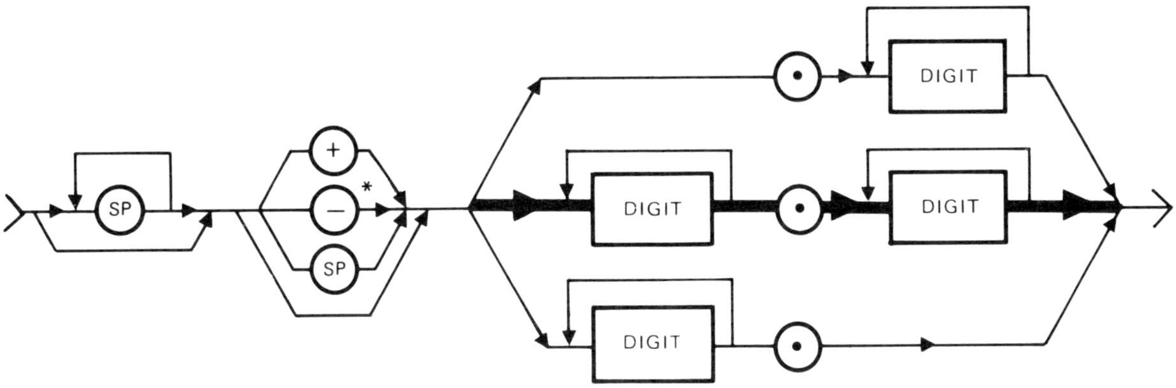

*Not preferred for ZERO valued numeric

Fig 9
NR2 Syntax Diagram

Numeric digits=
| 0 | 1 | 2 | 3 | 4 | 5 | 6 | 7 | 8 | 9 |

Special symbolic representations=
| E | SP | + | − | . |

The number of digits contained in the NR1, NR2, or NR3 notation is unspecified. *Within an NR field, the most significant digit (character) is sent first.*

The NR syntax diagrams and associated examples use the + symbol to indicate the positive signed value of a number. Typical devices and common usage frequently delete the generation of and ignore use of the | + | sign. Unsigned numeric data is therefore allowed and interpreted to represent a value equal to or greater than zero.

4.3.2.1.1 Numeric Representation Set 1 (NR1). NR1 consists of a set of implicit point representations of numeric values, that is, a radix point is implicitly considered to be placed (fixed but not transmitted) at the end of the string of digits. Both the unsigned and the signed representation may contain leading spaces. There shall be at least one digit and no embedded or trailing spaces allowed. The signed representation of the numeric value ZERO is preferred to have either a plus sign | + | or | SP | . NR1 representation is useful for conveying numeric data of integer values. The syntax for NR1 is shown in Fig 8.

4.3.2.1.2 Numeric Representation Set 2 (NR2). NR2 consists of a set of explicit point representations of numeric values with the radix point indicated by a decimal point | . | . For clarity the radix point should be preceded by at least one digit, possibly 0. The signed representation of the numeric value ZERO is preferred to have a plus sign | + | or | SP | . Embedded or trailing spaces are not allowed in NR2 representation. NR2 representation is useful for conveying numeric data that contains a fraction. The syntax for NR2 is shown in Fig 9.

4.3.2.1.3 Numeric Representation Set 3 (NR3). NR3 consists of a set of scaled representations with either implicit or explicit radix point together with an exponent notation as shown in Fig 10. Embedded or trailing spaces are not allowed in either the mantissa or exponent of NR3 representations. It is preferred that at least one digit, possibly ZERO, precede the radix point | . | .

4.3.2.1.4 Suffix. One special form of numeric representation (beyond the scope of ANSI X3.42-1975 [2]) useful in instrumentation systems permits use of a suffix following the numeric value whether the NR is represented in NR1, NR2, or NR3 form. The suffix is related closely to the NR in that it expresses the associated units: unscaled units preferred or with units multiplier if necessary. This suffix is expressed as a series of characters as shown in Fig 11. See Appendix C for typical suffix multipliers

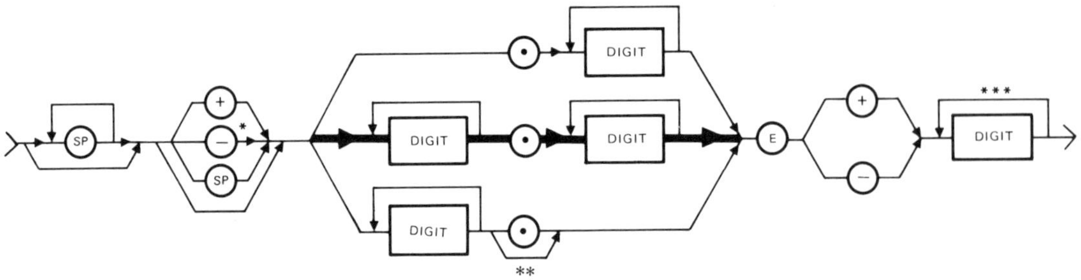

NOTES: (1) *Not preferred for ZERO valued numeric.
(2) **ANSI X3.42-1975 NR [2] notation allows for explicit-point-scaled representation only. Use of implicit point is allowed, not recommended for use with T or TE functions. All L, LE functions shall be able to handle both implicit and explicit forms. The implicit decimal point of the mantissa is assumed to be right-justified. L and LE functions shall accept unsigned NR3s as representing values equal to or greater than zero.
(3)***Two digits preferred, expressed in multiples of three (for example ...06, 03, 00, − 03..) for reasons of human interpretation and relationships to common physical units.
(4) NR3 representations that have an exponent with no sign should be accepted.
(5) It is permissible to send a zero value manitssa with a non-zero exponent (for example, 00000E+06). This representation is not preferred.

Fig 10
NR3 Syntax Diagram

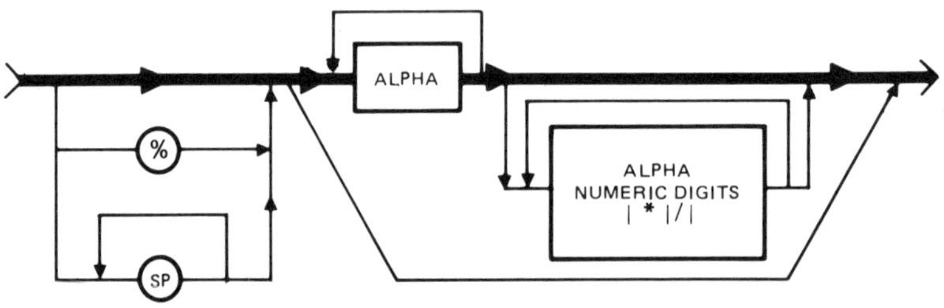

Fig 11
Suffix Syntax Diagram

Table 1
Numeric Representation Examples

Common Notation	0	0.00015	−10	−14.203	1234.5	5600
Signed NR1	+0	No Rep	−10	No Rep	No Rep	+5600
Unsigned NR1	0	No Rep	No Rep	No Rep	No Rep	5600
Signed NR2	+0	+0.00015	−10.0	−14.203	+1234.5	+5600.0
Unsigned NR2	0.	0.00015	No Rep	No Rep	1234.5	5600.0
Preferred NR2	0.0	0.00015	−10.0	−14.203	1234.5	5600.0
Signed R3	+0E+0	+1.5E−4	−1E+1	−1.4203E+1	+1.2345E+3	+5.6E+3
Unsigned NR3	0E+0	1.5E−4	No Rep	No Rep	1.2345E+3	5.6E+3
Preferred NR3	0.0E+00	150.0E−06	−10.0E+00	−14.203E+00	1234.5E+0	5.6E+03
With Suffix	0 V	0.00015 A	−10 V	−14.203 V	1.2345kHz	5.6kHz+

NOTES: (1) Although not used in the above examples, leading spaces and zeros are allowed, or both.
(2) The number of digits in these examples does not imply a preferred quantity of digits to be transferred.
(3) A ZERO value with a minus sign is not preferred.
(4) *No Rep* means no representation possible.

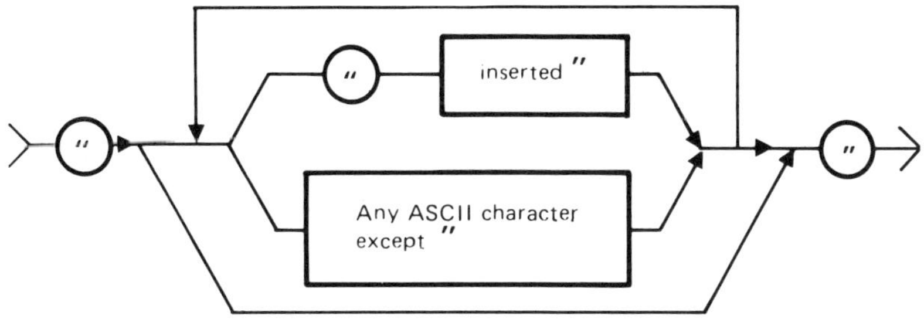

Fig 12
String Syntax Diagram

and units. See Appendix A for typical examples of the use of a suffix.

The non-preferred suffix path may be used to permit more human readable formating of data output to devices such as printers and CRTs (for example, 12.5% or 12.5 m/s).

4.3.2.2 String Data Field. The string data field allows ANY character in the ASCII 7-bit code (including non-printable characters) to be carried as a message. This data field is particularly useful where text is to be displayed (for example, on a printer or CRT type device). Using the string data type permits the use of format effectors such as | CR | | LF | SP | to correctly format text. Each string data field begins and ends with | " | . It is possible to include the quotation marks character | " | within the text by sending two sequential characters | "" | . The string data syntax is given in Fig 12.

4.3.2.3 Block Data Field. The block data field allows ANY 8-bit data type (including extended ASCII codes) to be carried as a message. It is particularly useful for sending large quantities of data. Since the focus of the block data field is to provide for long data streams, a means to increase data transfer reliability has been included via error detection.

A block data field is initiated by a *unique code*, the number | # | sign. The | # | *code is not allowed in any other* ANSI/IEEE Std 488-1978 [3] device dependent message.

EXCEPTIONS: (1) Within the string data field
(2) Within the block data field itself.

The block data field is used within the context of both the measurement and program messages as defined in Sections 5 and 6. In addition, the block data structure can exist as a separate entity (for example, it can stand alone for the transfer of files). It is also possible that an entire measurement or program message can be sent within the DABs of the block data field to take advantage of the error detection capabilities. The syntax for the block data field is given in Fig 13.

The data fields throughout this recommended practice are intended to provide common syntax structures for a broad range of block data types and, as such, represent a set of recommended general purpose solutions. More specialized applications requiring greater data field efficiency, shorter lengths or greater integrity, for example, may demand alternate block structures beyond the scope of this recommended practice.

Within the data field the most significant DAB shall be sent first.

Designers should consider the implications of buffer size and availability, particularly in the intended receiving devices. It is not to be assumed that all devices will automatically be capable of accepting lengthy messages.

Rationale for the selection of the appropriate error detection means is given in 4.3.2.3.3.

Throughout the balance of this recommended practice the block syntax diagram will be represented in a simplified form as expressed in Fig 14.

4.3.2.3.1 Block Preamble. The preamble consists of two bytes the first of which is the | # | byte. A second byte designates which of fourteen possible data types are to be represented within the block by the data bytes.

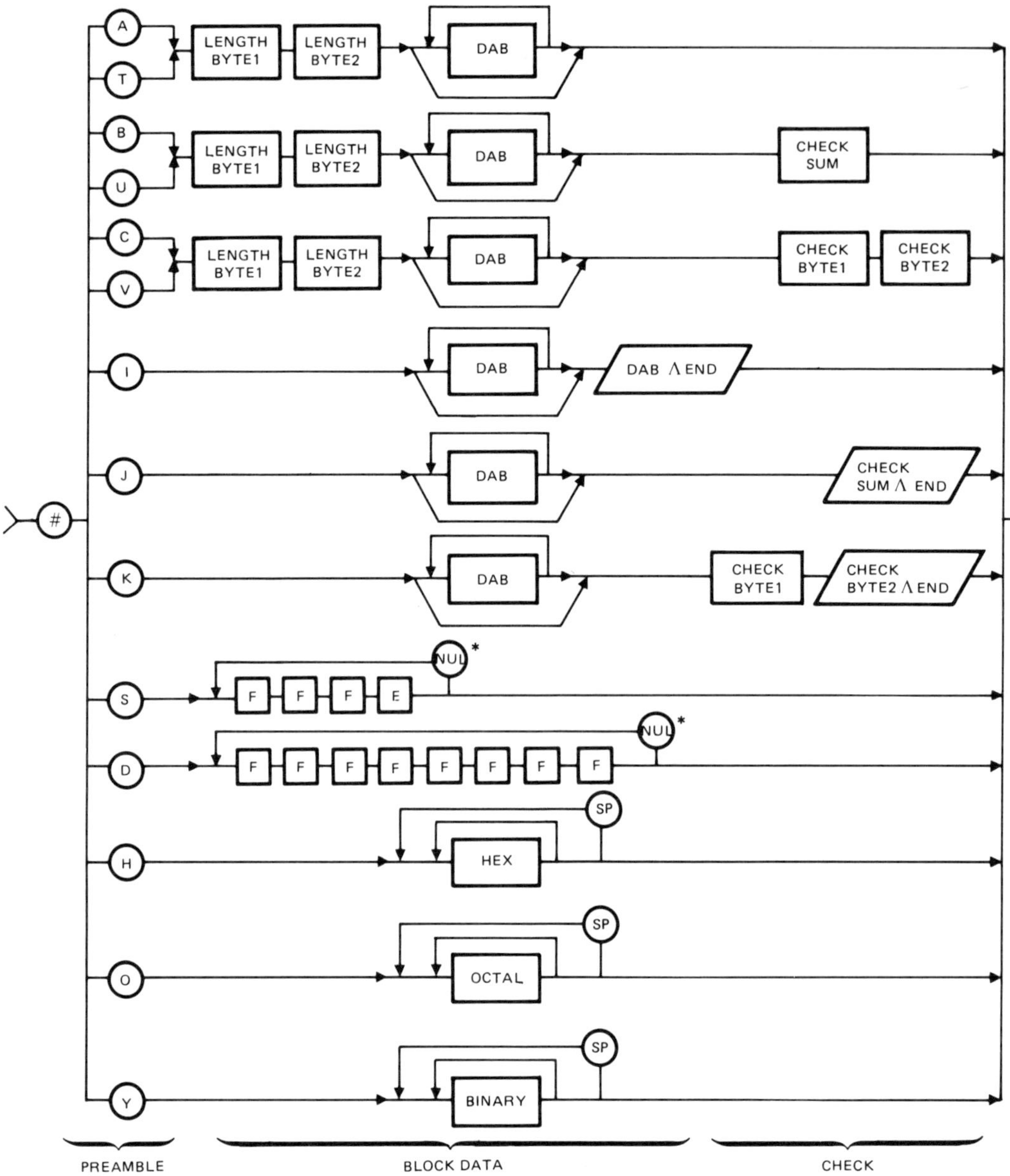

*NUL is not a unique code, therefore receivers may check every fifth, or ninth block data byte (for S and D representations respectively). If the byte is a "NUL" another floating point number will be sent/received. Some applications (for example, DMA transfers) may prefer multiple floating point numbers to be sent without "NUL" to simplify direct block transfers in which case the I format may be used.

Fig 13
Block Data Syntax Diagram

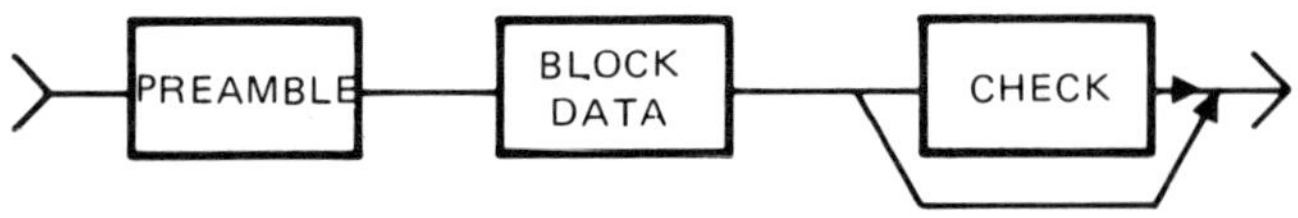

Fig 14
Simplified Block Data Syntax Diagram
(Used in Figs 21 and 23)

Table 2
Block Preamble Identifier Characters

<table>
<tr><th>Identifier</th><th>Check Type</th><th>Data Type</th><th>Usage</th></tr>
<tr><td>| A |</td><td>None</td><td>Binary</td><td rowspan="3">Data byte sequence (format) and coding is device dependent.</td></tr>
<tr><td>| B |</td><td>Sum</td><td>Binary</td></tr>
<tr><td>| C |</td><td>CRC</td><td>Binary</td></tr>
<tr><td>| T |</td><td>None</td><td>Text</td><td rowspan="3">ASCII DAB sequence is device dependent, bit 8 is a don't care bit, sent false if the application permits</td></tr>
<tr><td>| U |</td><td>Sum</td><td>Text</td></tr>
<tr><td>| V |</td><td>CRC</td><td>Text</td></tr>
<tr><td>| I |</td><td>None</td><td rowspan="3">Binary or Text</td><td>DABs that may be either binary or text and are of unspecified length
• END sent with last DAB</td></tr>
<tr><td>| J |</td><td>Sum</td><td>• END sent with checksum byte</td></tr>
<tr><td>| K |</td><td>CRC</td><td>• END sent with last check byte</td></tr>
<tr><td>| S |</td><td>None</td><td>Single Precision</td><td>DABs represent numbers expressed in single precision floating notation, 4 bytes per number (See Appendix E)</td></tr>
<tr><td>| D |</td><td>None</td><td>Double Precision</td><td>DABs represent numbers expressed in double precision floating point notation, 8 bytes per number (See Appendix E)</td></tr>
<tr><td>| H |</td><td>None</td><td>Hexadecimal</td><td>A sequence of DABs representing hexadecimal data</td></tr>
<tr><td>| O |</td><td>None</td><td>Octal</td><td>A sequence of DABs representing octal data</td></tr>
<tr><td>| Y |</td><td>None</td><td>Binary</td><td>A sequence of DABs representing binary data (coding different from | A | B | C | , see 10.5)</td></tr>
</table>

NOTE: No other identifier characters are allowed.

Preamble Check Type Coding

A or T or I — No error detection used.

B or U or J — Checksum detection means is used: The check byte contains modulo 256 check sum of the data bytes. [The length bytes are excluded from the checksum].

C or V or K — CRC16 Forward error detection is used: the CRC is performed on the data and check bytes. See 4.3.2.3.2.3 for CRC polynomial specifications. [The length bytes are excluded from the CRC].

4.3.2.3.2 Block Length Bytes. The length bytes contain information about the number of bytes in the block. The length byte bit assignments for L15 through L0 are given below and represent an unsigned binary integer that indicates the number of data bytes to follow.

Length Byte 1

DIO>	8	7	6	5	4	3	2	1
	L15	L14	L13	L12	L11	L10	L9	L8

Length Byte 2

DIO>	8	7	6	5	4	3	2	1
	L7	L6	L5	L4	L3	L2	L1	L0

L0 – L15 = Length count

L0 . . . L15 Coding
1,2,3 . . . 65,536

The integer decimal equivalent of these 16 bits specifies how many bytes are to follow. The count includes all data bytes and the check byte(s) if present. The use of values in the range of 0 to 4098 is preferred to facilitate file transfers between devices without requiring large data buffers.

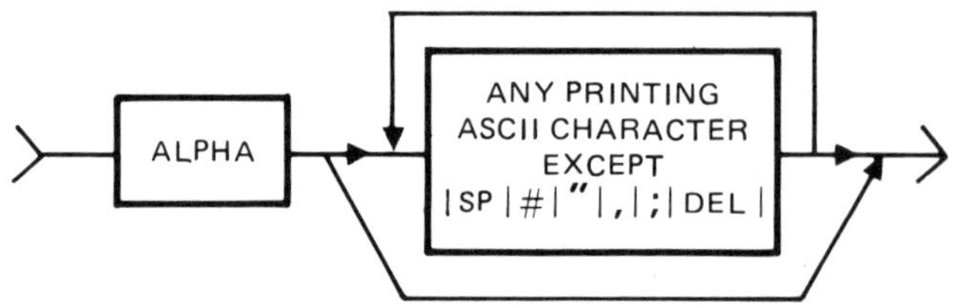

Fig 15
Character Syntax Diagram

4.3.2.3.3 Block Check Bytes. When a check byte(s) is used the check is performed on only the data bytes (DABS).

If a checksum error detection means is used, then the check byte shall contain the modulo 256 checksum.

If a CRC16 Forward error detection means is used, then the two byte count sent with the data bytes contains the remainder from division of the data byte stream by the CRC16 Forward polynomial (that is, $x^{16} + x^{15} + x^2 + 1$). The receiving device remainder is then zero for an error-free message. The parallel data on DI08 to DI01 shall be serialized most significant bit (DI08) first.

NOTE: The designer must ensure that the CRC circuitry can be enabled within 4 bytes times.

Rationale for the selection between the two detection means is as follows:

It is recommended that the CHECKSUM be used when short blocks of data are sent (for example, 15 bytes/block is a typical number). A checksum is insensitive to multiple errors resulting in a *no-error* indication from the checksum calculation. If the probability of a double error in the block is small, the user may choose the checksum method for single error detection.

It is recommended that the CRC16 polynomial method be used when the system is less sensitive to multiple errors yielding a *no-error* condition. Due to the nature of CRC16 technology the resultant error check is dependent to a significant degree on longer historical data patterns. The probability of multiple errors causing a *no-error* output of the error detection circuitry is less than the checksum method. The CRC16 approach is recommended where longer data blocks are being transferred. It is also recommended where the failure of error detection is critical and not to be tolerated.

4.3.2.4 Character Data Field. The character data field is used where words and text more clearly describe the nature of a program instruction than does a numeric data field. Character data fields always begin with an alpha character. The use of alpha characters only is preferred. Digits and other special purpose characters should be used only where the human readability and interpretation of the header is thereby enhanced. The character data syntax is shown in Fig 15.

4.3.3 Message Separators. A means to distinguish between one message unit and the next message unit or between information fields within message units is useful and sometimes essential for unambiguous message processing. Such means are useful for conveying related sets of data that occur in pairs (for example, amplitude and phase) or other data set multiples (for example, time, channel, frequency) or long continuous message streams.

Message separators may be needed before a talker function exits the TACs state and at certain times throughout a TACs condition. The context in which the separators are described assumes normal message traffic without controller initiated interruption of a message sequence.

Separators fall into three broad categories based on a hierarchical relationship to one another. Although several categories contain multiple choices for separator usage, the relative hierarchical relationship is retained. Separators are not mandatory for all syntax structures (see Sections 5 and 6) as it is possible to construct message formats with implicit separation between message elements.

Provisions have been made for both simple and complex parsers in receiving devices. Some parsers may be able to make the information field separation implicitly, that is, separator characters are not required. For implicit separation the receiver knows what type of information is expected based on preceding or following characters. Other parsers may require explicit use of

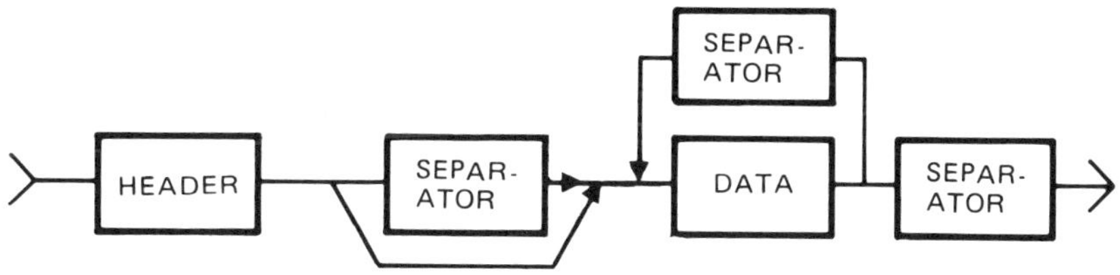

Fig 16
Typical Separator Use

any or all of the three categories (levels) of separators.

ANSI/IEEE Std 488-1978 [3] identifies the existence of two separator types without defining explicit usage, relationships to message formats, or coding. The END message ranks at the top of the hierarchy whereas the EOS message is at a lower level. This recommended practice defines three basic levels of separators for use with measurement and program message structures. The separators listed in descending order of the hierarchy are: SR3, SR2, SR1.

Figure 16 illustrates a very general structure for separator usage.

4.3.3.1 Separator Level 1 (SR1). An SR1 separator is typically used to identify the end of the lowest level of message element or data field(s). Two separators exist at this level: the comma ⌈ , ⌉ and semicolon ⌈ ; ⌉ . The lowest order separator of these two is the ⌈ , ⌉ and is, for most applications, the preferred separator. This is based on the fact that the comma ⌈ , ⌉ is of lower order than the ⌈ ; ⌉ in the written word. The syntax for SR1 is given in Fig 17.

4.3.3.2 Separator Level 2 (SR2). An SR2 separator is typically used to separate a sequence of message units at a distinctly higher level than that of the SR1 separators. There is only one level of separator at the SR2 level although there are several implementation choices. The syntax of SR2 is given in Fig 18.

4.3.3.3 Separator Level 3 (SR3). The SR3 separator is the highest order separator specified by this recommended practice. An SR3 separator is usually used when one or a series of measurement or program messages has been completed. As the highest order separator, the SR3 or END message has special significance. A talker, having sent END, shall not output further DABs automatically. The *talker* must then receive a device-dependent message or a specific interface message (for example, GET) prior to resuming output. Systems using the local ton (talk only) message may resume output via an external stimulus such as *manual trigger*. The SR3 syntax is given in Fig 19.

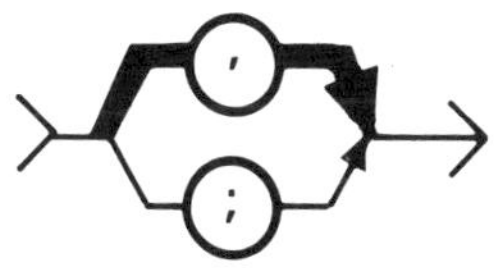

Fig 17
SR1 Syntax Diagram

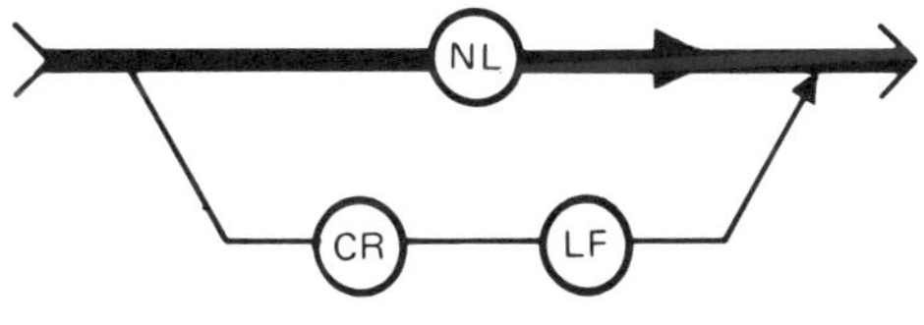

NOTES: (1) ASCII code for NL is identical to LF.

(2) The ability to select between the SR2 separator paths within a device will provide more universal information interchange

(3) FUTURE: The long term intent is to use a SINGLE character for the SR2 separators. Use of the double CRLF character separator (alternate path) within a device may be needed for some time to facilitate compatibility with existing products.

(4) PAST/PRESENT: For broad compatibility the syntax path for the T (TE) function has been/is CRLF whereas the preferred path for the L (LE) function is LF or NL. The L function should use LF as the separator and NOT the CR character. The CR character should be discarded unless the device (like a printer) is capable of executing the CR operation.

Fig 18
SR2 Syntax Diagram

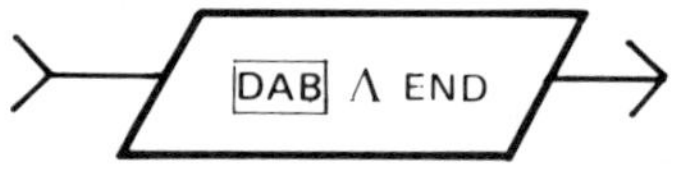

NOTES: (1) The END message (sent on the EOI signal line) is sent concurrent with DAB and follows identical timing requirements (see 2.3.3.4 of ANSI/IEEE Std 488-1978). [3]

(2) DAB is a device-dependent multiline message encoded on DI01-DI08. DAB may be a normal data byte or one of the other lower order separators as specified in subsequent diagrams.

**Fig 19
SR3 Syntax Diagram**

4.3.3.4 Leaving and Entering TACS/LACS. The use of SR1, SR2, and SR3 separators shall not imply that a given device will automatically exit the TACS or LACS states.

Any data transfer may be interrupted synchronously by the controller function to achieve other higher priority operational sequences. It is recommended that the interrupted device resume data byte transfer with the next byte following the byte transmitted prior to the point of interruption (data transfer resumes when the device re-enters TACS). It is recommended that asynchronous interruption of a message be avoided as it is difficult (if not impossible) to determine whether the byte being transferred at the time of interruption was received.

4.3.3.5 Separator Coding. The allowable codes for separators are given in Table 3. This is a summary coding table representing a composite of the SR capabilities defined throughout 4.3.4.

**Table 3
Separator Coding Summary**

Separator	Code	Remarks
SR1 (Level 1)	, ;	See Note 1
SR2 (Level 2)	NL CRLF	See Notes 2,3
SR3 (Level 3)	END	Sent on EOI, See NOTE 3

NOTES: (1) To facilitate explicit parsing between the header and data fields in some applications an exception to the above SR1 coding schema is required. For a restricted set of conditions one may use the SP header separator between header and data fields. See Fig 24 for details.

(2) The NL (new line) or LF (line feed) is preferred over the CRLF code pair as the SR2 separator. Use of multiple bytes to perform the separator function may lead to an ambiguous response in some devices. Therefore, use of the single byte NL (ASCII code column 0 row 10, same code as LF) is preferred to perform the SR2 separator function. All these codes are format effectors*.

(3) The ETB and ETX codes defined in the ASCII code may be found as record separator and terminator respectively in some international use of the related IEC Publication 625-2. The ETB and ETX codes are ASCII communication control* characters and therefore should be used with caution. Neither code is recommended as a means to effect separator action.

*Refers to ANSI X3.4–1977 [1]

5. Measurement Messages

Instrumentation devices perform many measurement functions. Typical tasks measuring frequency or voltage, digitizing a waveform, and performing network analysis result in the generation of measurement results. The specific content of the message is device-dependent, however the organization or general format of these application dependent messages can be structured in a way common to many devices. The purpose of this section is to establish a set of recommended practices for the format of measurement messages such that information interchange among independently designed devices is achieved at the highest level possible.

5.1 Measurement Message Overview. The individual syntax diagrams defined throughout Section 4 may be used to construct and assemble specific measurement messages suited to unique applications. A generalized form of typical measurement messages is given in Fig 20 to introduce the normal flow of data fields, the *format* of the message. The reader is cautioned not to design to this daigram; it serves only to introduce the general flow of information.

The combination of a header and the measured data itself form what is termed the measurement result. This is equivalent to a single message unit. Some form of separator is then used to distinguish this result from the next one. The optional nature of the header is denoted by the bypass path around the header element.

NOTE: For measurement messages the data and separator fields are mandatory.

5.2 Measurement Message Formats. The explicit syntax for a common set of measurement messages in a way considered to be relatively device and application independent is set forth here. The specific content of each message element, at the lowest level (for example, semantic meaning of the letter P), is beyond the scope of this recommended practice. Figure 21(a) sets

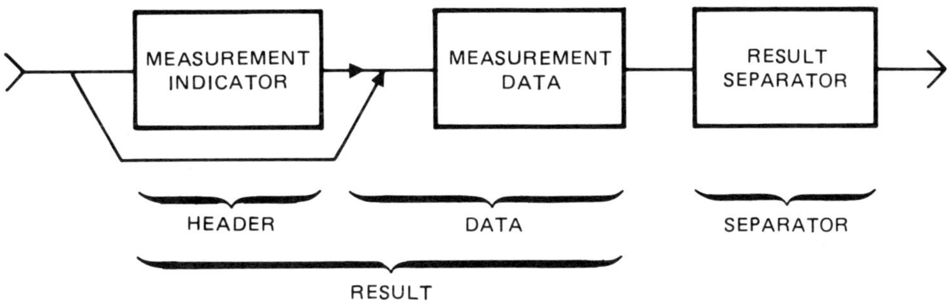

Fig 20
General Form: Measurement Message

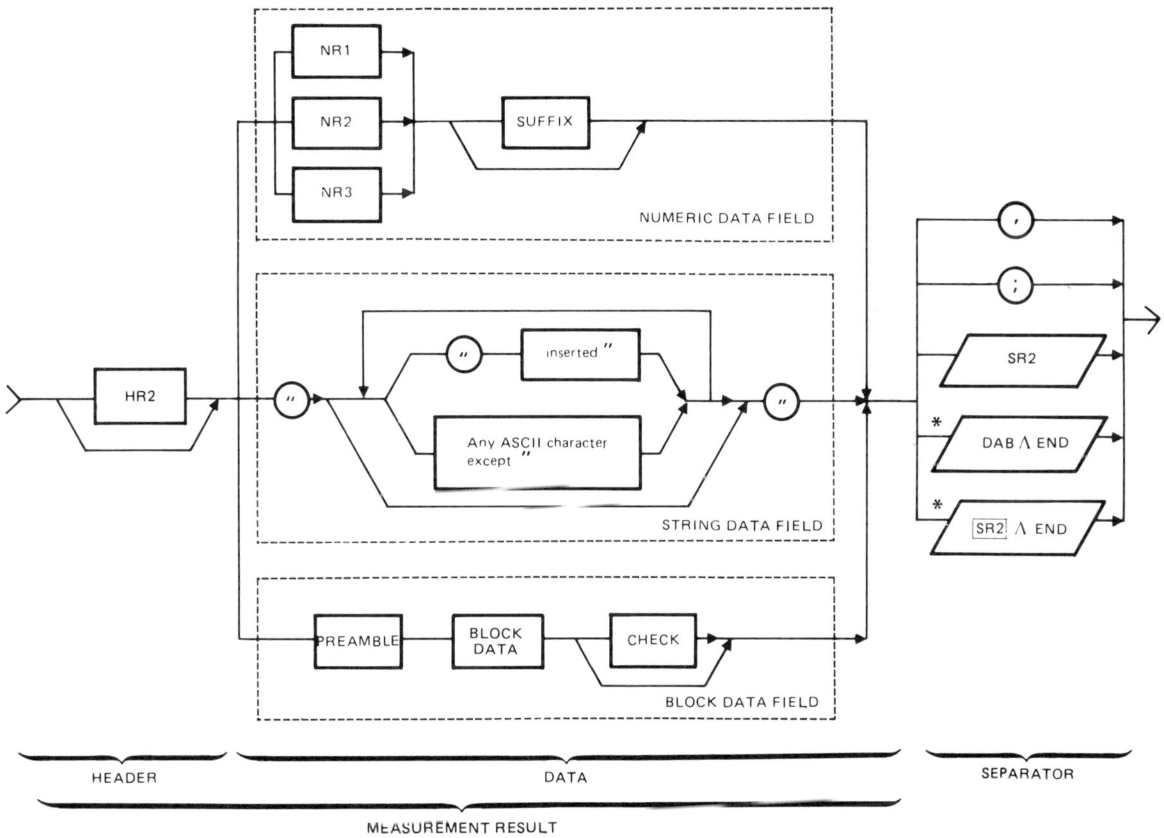

*If check field was END message, then omit separator field.

Fig 21
(a) Measurement Message Syntax Diagram

forth the preferred syntactic construction of the three different types of measurement results identified as numeric data, string data, and block data. Figure 21(a) is to be viewed as a *single* measurement message, one result. Many real applications, of course, require a series of results. Figure 21(b) will indicate the method(s) available to achieve a variety of more complex messages.

5.2.1 Measurement Message Construction. These guidelines and recommendations are provided to facilitate the construction of measurement messages intended for the highest level of interchange possible between independently designed devices.

5.2.1.1 General. To accommodate a wide variety of measurement devices the measurement message may contain any number of DABs.

DABs are sent the most significant DAB first for numeric data.

5.2.1.2. Header Field

5.2.1.2.1 The preferred HR field should contain only alpha characters. This facilitates reading of the NR field with free-field formats in common use. The HR field is particularly useful where human readibility is important.

5.2.1.2.2 The HR2 field is optional and, if used, should be as short as possible. For a known product, in a specific mode, a fixed length HR2 field greatly facilitates parsing of the data fields by the listener. All characters, including embedded spaces, make up the HR field. Spaces may be used to maintain a fixed length HR.

5.2.1.2.3 The HR2 field, when used to specify units associated with the NR data field, shall use unscaled units as the preferred set (for example, volts, ohms, amperes). If scaled units are used then the preferred values should be in multiples of 10^{+3} or 10^{-3} (for example, μF, μV, kA, mm). The NR suffix field may, in addition, be used to specify the units and multiplier relating to the numeric data field in which case unscaled units are preferred.

5.2.1.2.4 If the HR2 field refers to data quality, then device-dependent alpha characters may be used to indicate specific conditions. For example, calibration voltage may be represented by CV.

5.2.1.3 Data Field. It is *mandatory* that at least one of the data fields be used in a given measurement message.

5.2.1.3.1 Numberic Data.

(1) The NR field shall contain either an NR1, NR2, or NR3 value. The choice is application dependent.

(2) NR1 implicit point representation is most useful for numeric data where either limited interpretation or generation of data is required (for example, restricted or fixed-range) and also where large volumes of fixed format data or high speed data are transferred between devices. In these applications data rates are of primary significance.

(3) NR2 explicit-point unscaled representation is most useful for numeric data where either the range of data output is limited or data is intended to be used with devices where human interpretation is prevalent.

(4) NR3 exponential notation is preferred where measurement or controller devices must accommodate a wide range of values or where the specific range of data to be output (or input) is unpredictable.

(5) A constant length NR field length for a specific operating mode may facilitate efficient data interpretation and storage. For a given operating mode the selected type of NR representation should remain the same, even through ZERO values.

5.2.1.3.2 String Data.

(1) The string data type is intended where the message is to be displayed for human interpretation as on a printer, plotter, or display unit.

(2) The ability to carry any ASCII 7-bit character enables the message to contain nonprinting characters and thus messages can be readily formatted for ease of human interpretation.

5.2.1.3.3 Block Data.

(1) Block data is intended to be used where the message is of extended length.

(2) Block data supports relatively transparent data transmission and thus provides for the capability of sending binary data messages.

(3) To increase the data integrity for long message lengths, several alternate check means are provided. The rationale for the use of each check method is given in 4.3.2.3.3.

(4) Each block data field requires a mandatory preamble consisting of the | # | flag followed by an alpha flag. The block data set is limited to seven explicit choices:

(a) | A | B | C | T | U | V fields contain explicit length and check mechanisms for binary and text data

(b) | I | J | K | fields contain explicit check mechanism for binary or text data of indefinite length

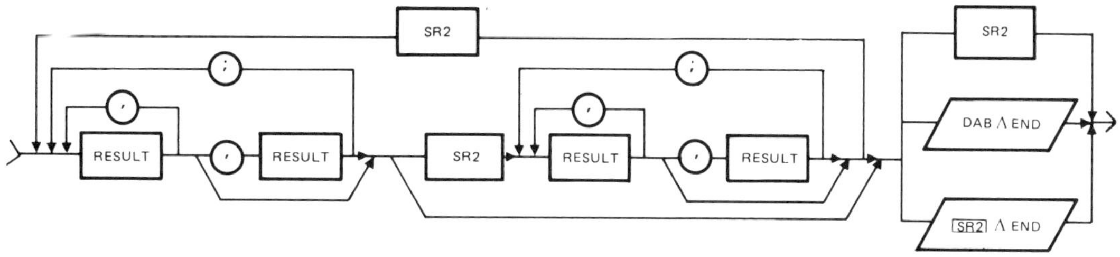

Fig 21
(b) Separator Usage and Hierarchy Rules

(c) | S | D | fields support floating point numbers, have neither length nor check mechanisms and use the | NULL | code to separate multiple values

(d) | H | O | Y | fields are used to represent non-decimal data with a selected set of ASCII codes

(5) Designers should consider the impact of buffer size and availability and not assume that recipients of long records will be capable of accepting such records automatically.

5.2.1.4 Separator Field.

(1) Each result shall be terminated with one of the separators identified in Fig 21(a).

(2) Guidelines for the selection of a specific separator are given in Fig 21(b) to accommodate the construction of more complex messages containing more than one result.

(3) Use of the END message indicates the device expects to be further instructed (that is, receive a device-dependent message or an interface message such as GET) before another measurement data message is output. Systems using the local ton (talk only) message may resume output via an external stimulus such as *manual trigger*.

5.2.2 Alternate Measurement Message Construction. This recommended practice does not define the mixing of program and measurement message units. The occasion may arise when measurement results from one device need to be transmitted for *processing* to another device. In this case measurement result(s) may be formatted according to the program message syntax constructs as defined throughout Section 6.

6. Program Messages

Instrumentation devices are called upon to perform many different tasks. This process of performing different tasks requires changes in the basic configuration of the device in preparation for performing the basic measurement or stimulus task. A device might be required to change operating mode, measurement range, or output mode, for example. Receipt of program instructions effect the necessary changes. The specific content of this data is device-dependent, however the organization or general format of these application dependent messages can be structured in a way common to many devices. The purpose of this section is to establish a set of recommended practices for the format of program messages such that information interchange among independently designed devices is achieved at the highest possible level.

6.1 Program Message Overview. The individual syntax diagrams defined throughout Section 4 may be used to construct and assemble specific program data messages suited to unique applications. A generalized form of typical program messages is given in Fig 22 to introduce the normal flow of data fields, the *format* of the message. The reader is cautioned not to design to this diagram, it serves only to introduce the general flow of information.

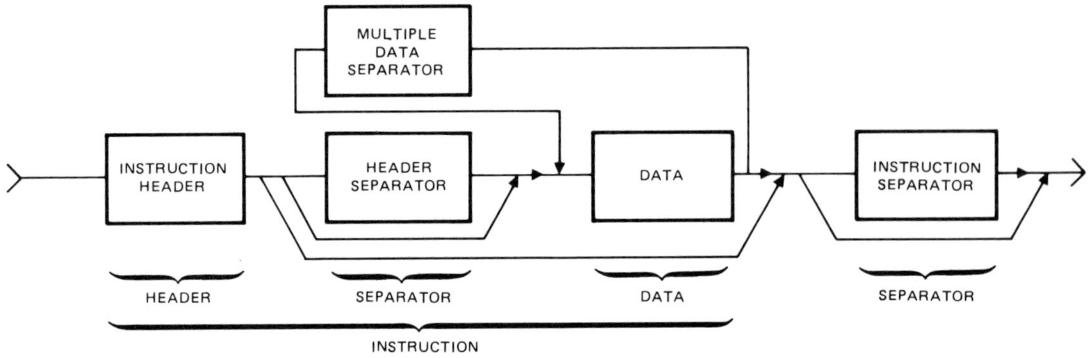

Fig 22
General form: Program Message

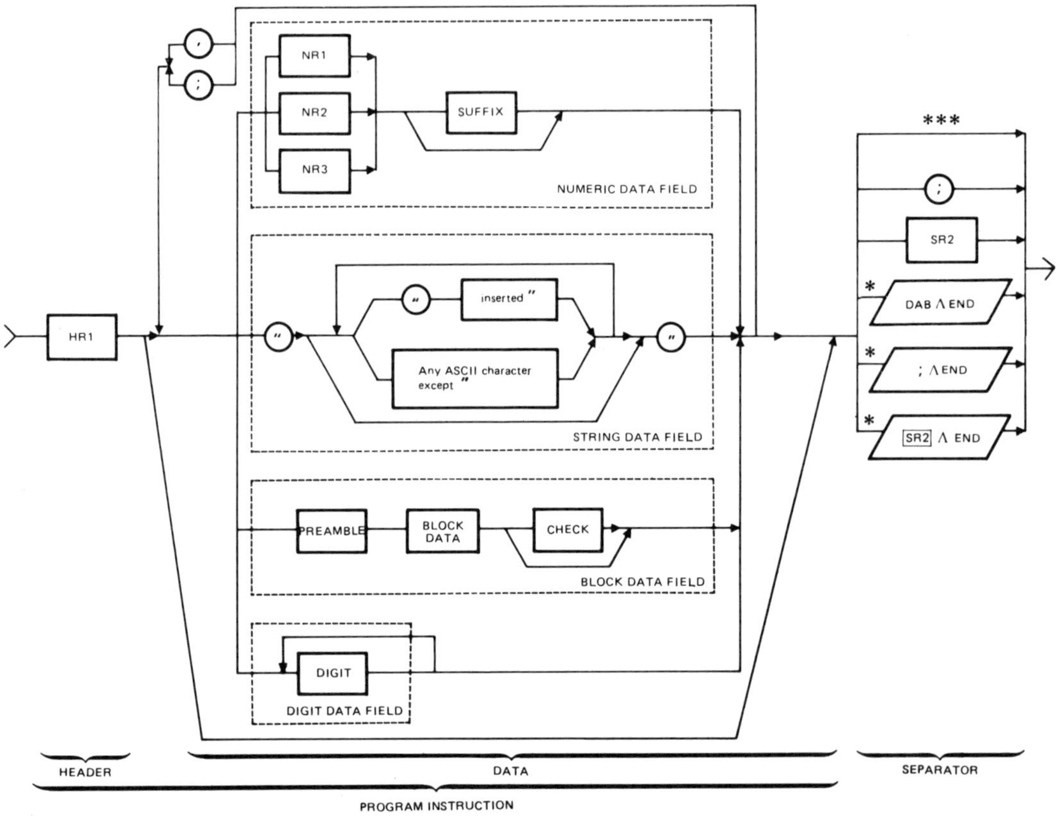

*If check field uses END message, then omit separator field.

***WARNING: Any data field ending with an alpha character requires an unambiguous means of separation.

Fig 23
Program Message Syntax Diagram

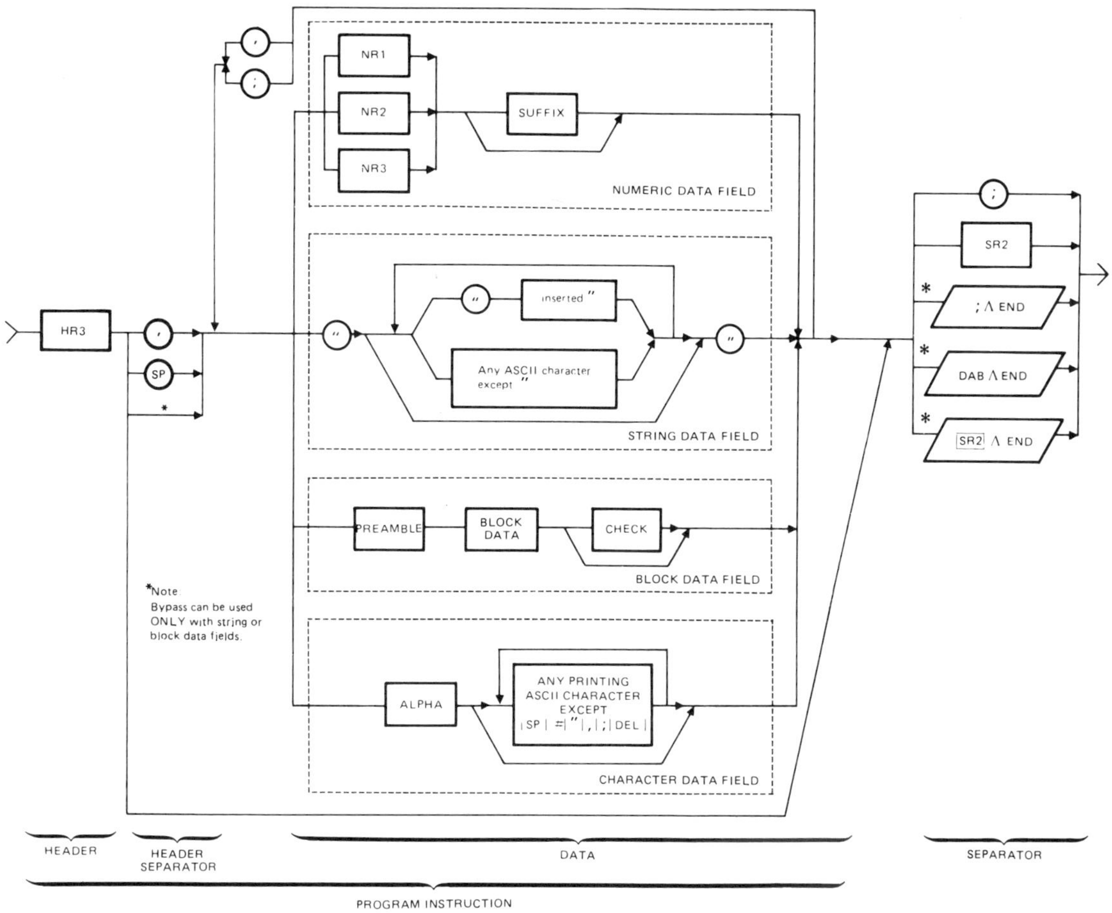

*If check field was END message, then omit separator field.

Fig 24
Program Message Syntax Diagram

The combination of a header, under special conditions a separator, and the instruction data itself form what is termed a program instruction. This is equivalent to a single message unit: the instruction is processed as a unit. Some form of separator is then used to distinguish this instruction from the next one. The instruction header field for program messages is mandatory. The optional header separator/data pair is denoted by the bypass path around these elements.

6.2 Program Message Formats. The explicit syntax for a common set of program messages in a way considered to be relatively device and application independent is set forth here. The specific content of each data field, at the lowest level (for example, semantic meaning of the letter P), is beyond the scope of this recommended practice. Figs 23 and 24 set forth the preferred syntactic construction of four different types of program instructions identified as digit data, numeric data, string data, and block data. Either Fig 23 or Fig 24 are to be viewed as a SINGLE program message or instruction. Many real applications, of course, require a series of program instructions. A subsequent diagram will indicate the method(s) available to achieve a variety of more complex messages.

The syntax shown in Fig 23 is simpler than that shown in Fig 24. A header separator is not used and it is possible to bypass the message separator. This is made possible by limiting the header to the alpha (HR1) characters only. The main focus of this type of syntax diagram is to facilitate machine-to-machine communication and to facilitate simplicity in the programming of devices. The latter can be achieved by a high degree of correlation between front panel control markings and the content of program messages. The resultant instructions are more compact and less human readable than those produced by Fig 24.

Each data field in Fig 24 is explicitly separated. This permits the content of the header field to contain non-alpha characters. Likewise the data field can now incorporate a character data field. Use of the space as a header delimiter permits the generation of messages more oriented toward human readability. The use of special characters within the header and the character data field permits duplication of similar characters located on the front panel (for example, VOLTS/DIV, TRIG, AUTO). Possible similarities between program instruction and actual instrument operation can facilitate a self-documenting program. The more human-like instructions derived from Fig 24 will require, in general, more sophisticated parsers.

6.2.1 Program Message Construction. These guidelines and recommendations are provided to facilitate the construction of program messages intended for the highest level of interchange possible between independently designed devices.

6.2.1.1 General.

(1) To accommodate a wide variety of devices the program message may contain any number of DABs.

(2) The DABs are sent most significant DAB first for numeric data.

(3) Program messages are normally input to a device. At times it is useful to output a program message for storage in some external device and subsequent recall to the originating device. Figures 23 and 24 may be considered as useable in this operational mode.

6.2.1.2 Header Field. It is *mandatory* that one of the header data fields be used in each specific program instruction.

For Fig 23, the HR field should be as short as possible, contain alpha characters only, and shall be unique for each individual message function.

6.2.1.3 Data Field

6.2.1.3.1 Numeric Data. Use of the alpha suffix permits an instruction to be written and read in normal engineering units. For example, the message CF123.45 kHz establishes a center frequency of 123.45 kHz where CF is the header and kHz the suffix.

6.2.1.3.2 String Data.

(1) String data is intended where the message is to be displayed for human interpretation as on a printer, plotter, or display unit.

(2) The ability to carry any ASCII 7-bit character enables the message to contain non-printing characters and thus messages can be readily formatted for ease of human interpretation.

EXAMPLE: "CONNECT PROBE TO TEST POINT #1"

6.2.1.3.3 Block Data.

(1) Block data fields are intended to be used where the message is of extended length.

(2) Block data supports relatively transparent data transmission and thus provides for the capability of sending binary data messages.

(3) To increase the data integrity for long message lengths, two alternate check means are provided. The rationale for the use of each check method is given in 4.3.2.3.3.

EXAMPLE: #S 01000010 11100000 00000000 00000000
(These four bytes in binary notation following the preamble represent the number 1.12E+02 in single floating point notation, see Appendix E)

(4) Designers should consider the impact of buffer size and availability and not assume that recipients of long records will be capable of accepting such records automatically.

6.2.1.3.4 Character Data. The character data field permits the use of human oriented instructions. It is allowed only with Fig 24.

EXAMPLE: DISPLAY ON or DISPLAY OFF where DISPLAY is the header, ON and OFF are character data fields.

6.2.1.3.5 Digit Data.

(1) Use of an HR1 field followed by one or more digits provides an abbreviated way to program a specific function (for example, range R3). This method of programming allows equivalent front panel switch positions (for example, high, medium, low) to be selected where precise numeric values are not required.

EXAMPLE: R2L1
This example illustrates a simplified program instruction for *range 2* (10 volts), *level 1* (level off)

(2) Allowed only in Fig 23.

6.2.1.4 Separator Fields. Figures 25(a), 25(b), and 25(c) provide definitions for three basic program instruction formats. Their purpose is to specify the manner in which more complex program instruction messages and sequences can be constructed and organized. Each diagram defines an orderly and consistent relationship among the several available separator types. Each indi-

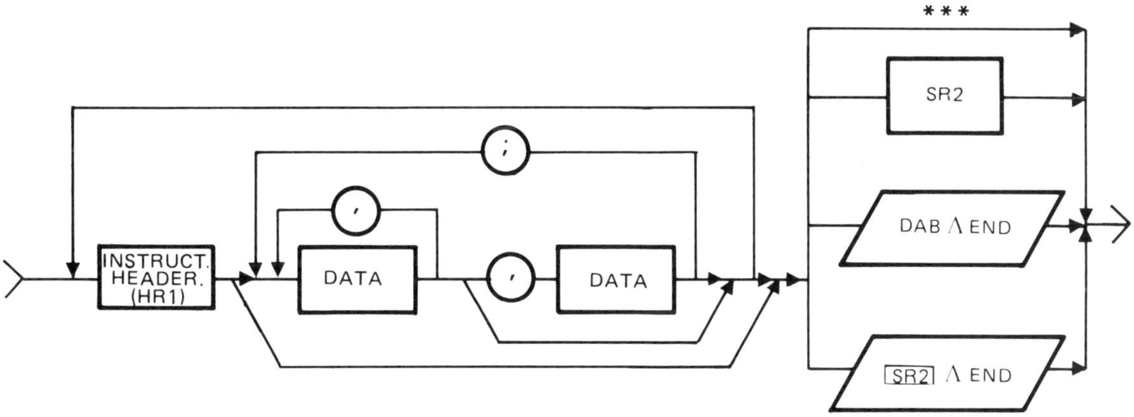

(a)

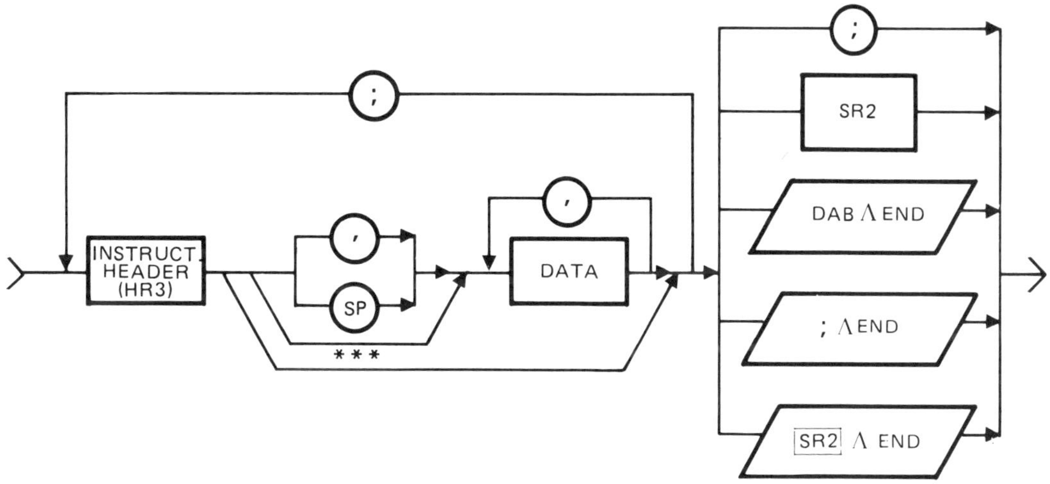

(b)

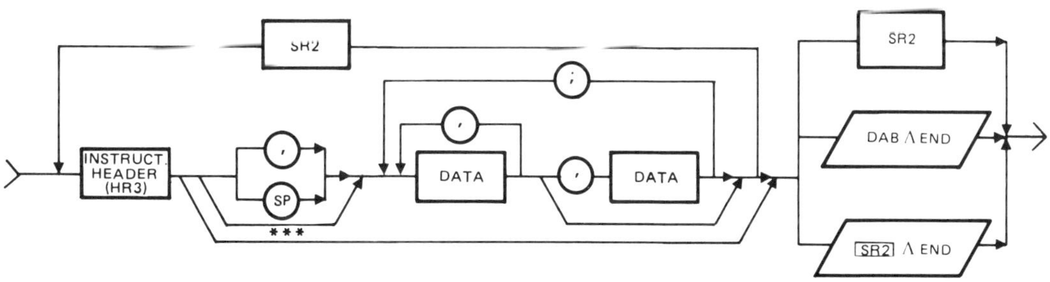

(c)

*** **WARNING:** Any data field ending with an alpha character requires unambiguous means of separation (for example set, cal, test).

Fig 25
(a) Separator Usage and Hierarchy Rules (Applies to Fig 23 Only)
(b), (c) Separator Usage and Hierarchy Rules (Applies to Fig 24 only)

vidual syntax diagram may, in turn, be re-entered by looping back from rightmost exit to leftmost entrance with one exception. When an END message separator (that is, SR3) is encountered the diagram is no longer re-entrant. The three syntax diagrams of Figs 25(a), 25(b) and 25(c) provide substantial flexibility for the construction of program instructions.

It is recommended that a device's hierarchical separator capabilities be explicitly defined in the relevant product documentation to facilitate use of the product in system configurations. Appendix B provides further guidance in this matter.

Specific requirements and guidelines are as follows:

(1) Each instruction shall be terminated with one of the separators identified in either Fig 23 or Fig 24.

(2) It is preferred that, within a device, only ONE hierarchical diagram be used (that is, Fig 25(a), or 25(b), or 25(c)).

(3) To achieve the more English-like instruction capability defined in Fig 24, the | SP | is used as a special header separator in Fig 25(c). The reader is cautioned to consider the impact of this implementation as the use of | SP | no longer has a universal meaning. A given product in a specific operating mode shall use only one of the header separators (that is, | , | or | SP |).

7. Status Message

ANSI/IEEE Std 488-1978[3] defines both serial and parallel means for sending status messages. This section defines further the structure of a single status data byte sent during a serial poll. Parallel poll message structure is specified in the standard.

A status message may be sent from a device with an STB message (DI01...6, DI08) in response to a serial poll sequence when the device is in the SPAS state. The principle purpose of the STB message is to present critical summary and detailed status to the controller-in-charge. Summary status means the logical OR or detailed status within the same category, for example, if more than one abnormal condition exists. A wide variety of device-dependent internal states and conditions may be carried by STB messages and therefore a complete data structure code assignment is not feasible. A degree of compatibility between devices can be accommodated when the lines DI01...6 are used according to the conventions in this section. The RQS message (DI07) is sent concurrent with the STB message.

7.1 Summary Status Conditions. DI06 sent true reports a summary status condition associated with abnormal operation of a device. Typical use would include internal error conditions within the device functions, erroneous program message sent to a device, incomplete or erroneous measurement message, limit or alarm conditions. DI06 sent false reports normal operation is occurring or has been completed (for example, currents within limits).

7.2 Busy/Ready Condition.

NOTE: At one time it was considered appropriate to reserve DI05 to indicate the fact that a device was either *busy* completing its assigned task or *ready* to start another one. In this case, the *ready* message could indicate either ready to have *data read out* or ready to *start new operation*. The meaning of *ready* could be ambiguous. The demand for additional device dependent status message code space led to the recommendation of 7.3. If it is meaningful to indicate busy/ready messages, then use of DI05 is the appropriate signal line. IEC Publication 625-2 specifies DI05 as a busy/ready data bit.

Table 4
Status Message Structure and Code Assignment

Messages	RQS	STB			
Logical Value	DI07	DI08	DI06	DI05	DI04—DI01
1	Requested Service	X	Abnormal	X	X X X X
0	Not Requested Service	X	Normal	X	X X X X

X = Device-dependent code assignment

7.3 Additional Conditions. DI01-5 and DI08 are available for designer assignment to report either additional summary status or more detailed status. The six bits (DI01-DI05 and DI08) may be used in any manner to report device-dependent conditions.

7.4 General Guidelines. The preferred code assignment for the STB messages are given in Table 4.

The content of the STB message sent on DI01-6 is free to change between STB message transfers as the internal device functions change. However, the coding of the messages for a given device is not permitted to change. Status message as used herein refers to a single STB message byte as defined by ANSI/IEEE Std 488-1978[3]. If more detailed status data is required, then device-dependent means are to be used.

The conventions of this section are intended to be used where device-dependent requirements do not dictate otherwise.

If a device has one and only one reason for requesting service, then the status for this reason may be indicated on the RQS bit.

8. Display Messages

This recommended practice does not contain a specific recommendation for the syntax of data uniquely utilized for display purposes. Both the program and measurement messages, however, can be used for display purposes. The character and string data types are particularly useful for this purpose for which the specifications in Sections 5 and 6 are considered useful.

Display messages are intended for device input or output where human interpretation is important. Display messages should follow Section 5 specifications for measurement messages as far as possible.

9. Data Shift Techniques

Some applications require that DABs, data bytes input to a device, be interpreted as one type of data during a given series of message unit transfers and as another type of data during subsequent message unit transfers. Sections 5 and 6 of this recommended practice clearly identify several different data types. The explicit means of intermixing different data types (for example, program instructions and measurement results) is beyond the scope of this recommended practice. However, 5.2.2 gives some general guidelines.

ANSI/IEEE Std 488-1978[3], 5.5 indicates one available resource, multiple listen addresses, to facilitate a given device's reception of two (or more) different data types.

The designer may utilize unique device-dependent coding to select between different data types while that device is a single addressed listener. Care must be exercised to avoid ambiguous code assignments. Assignment of these special *shift* codes may reduce the remaining code space available for measurement, program, or display data.

10. Data Coding

10.1 Recommended Code. ANSI X3.4-1977 [1] ASCII 7-bit code is the recommended data representation code for communication of device dependent messages as described throughout Sections 5 and 6. The ASCII 7-bit code is shown in Table 5.

10.2 Hexadecimal Code. The ASCII 7-bit code set may be used to represent hexadecimal data for communication of selected device dependent messages. The code set shall be as follows:

| 0 | 1 | 2 | 3 | 4 | 5 | 6 | 7 | 8 | 9 | A | B | C | D | E | F |

10.3 Octal Code. The ASCII 7-bit code set may be used to represent octal data for communication of selected device dependent messages. The code set shall be as follows:

| 0 | 1 | 2 | 3 | 4 | 5 | 6 | 7 |

10.4 Floating Point Coding. The coding techniques for floating point are described in Appendix E.

10.5 Binary Code. The ASCII 7-bit code set may be used to represent binary data for communication of selected device dependent messages. The code set shall be as follows:

| 0 | 1 | X |

Table 5
ASCII 7-bit Code

Bits b7 b6 b5 →					0 0 0	0 0 1	0 1 0	0 1 1	1 0 0	1 0 1	1 1 0	1 1 1
b4	b3	b2	b1	Column → / Row ↓	0	1	2	3	4	5	6	7
0	0	0	0	0	NUL	DLE	SP	0	@	P	`	p
0	0	0	1	1	SOH	DC1	!	1	A	Q	a	q
0	0	1	0	2	STX	DC2	"	2	B	R	b	r
0	0	1	1	3	ETX	DC3	#	3	C	S	c	s
0	1	0	0	4	EOT	DC4	$	4	D	T	d	t
0	1	0	1	5	ENQ	NAK	%	5	E	U	e	u
0	1	1	0	6	ACK	SYN	&	6	F	V	f	v
0	1	1	1	7	BEL	ETB	'	7	G	W	g	w
1	0	0	0	8	BS	CAN	(	8	H	X	h	x
1	0	0	1	9	HT	EM	)	9	I	Y	i	y
1	0	1	0	10	LF	SUB	*	:	J	Z	j	z
1	0	1	1	11	VT	ESC	+	;	K	[	k	{
1	1	0	0	12	FF	FS	,	<	L	\	l	|
1	1	0	1	13	CR	GS	-	=	M	]	m	}
1	1	1	0	14	SO	RS	.	>	N	^	n	~
1	1	1	1	15	SI	US	/	?	O	___	o	DEL

(NOTE: NL code is Identical to LF Code)

11. Code Elements Related to Signal-Lines

11.1 General Coding Rule. In general, the least significant bit of a multi-bit code transferred concurrently across the interface is placed on the DI0 line with the lowest number. An 8-bit binary code shall utilize DI01 through DI08 to represent bits 2^0 through 2^7. It is preferred that for a single data byte the data byte be right justified (for example, a 5-bit code utilizes DI01 through DI05 respectively to represent bits 1 through 5). Unused signal lines should be set false.

11.2 ASCII 7-Bit Code Element Relationships. The ASCII 7-bit code bits shall be assigned to the DI0 signal lines as shown in Table 6. DI08 is set false if it is not used for any specific purpose (for example, DAB or parity).

Table 6
Code/Line Relationship

Column of ASCII 7-bit Code		b7	b6	b5	b4	b3	b2	b1
DIO line	DI08	DI07	DI06	DI05	DI04	DI03	DI02	DI01

12. Error Detection

The need for error detection capability within instrumentation systems varies significantly in relation to the nature of the noise environment, nature and importance of data carried on the interface, type of device function active at both data source and acceptor, and the overall system application.

The choice of which error detection technique should be used is application dependent, therefore a single method is not recommended. Several common techniques are as listed in 12.1 and 12.2, along with comments about their application.

12.1 Error Detection "at the Bit Level"

12.1.1 A simple lateral parity bit on DI08 to delect errors contained in DI01-7 for the ASCII 7-bit code provides minimal means for error detection and requires minimal hardware. Parity check permits detection of a single error within the bit grouping of a byte. Multiple-bit errors within a byte may not be detected.

Use of DI08 as a parity bit is not preferred.

12.1.2 A longitudinal parity check bit on any given DI0 line at the end of a data record may be used in the same way as the lateral parity check bit for the same purpose and results. See 4.3.2.3.3 for guidance on the use of checksum techniques.

12.1.3 A cyclic redundancy check (CRC) is much more comprehensive and assures a higher degree of error detection capability. See 4.3.2.3.3 for CRC16 for details.

12.2 Error Detection "at the Message Level"

12.2.1 An instrument receiving a programming command message can check the message for proper syntax so that all the data that is expected has the proper format. Commands not having the proper syntax should not be executed. Instead, a service request (SRQ) message should be generated and the status byte (STB) on the subsequent serial poll should indicate a syntax error.

12.2.2 An instrument receiving a programming command message can check the semantics of the message to see if they are meaningful and executable. For example, a power supply with a maximum output of 50 V should not attempt to execute a command that tells it to output 500 V. Instead, a service request should be generated and the status byte on the subsequent serial poll should indicate an execution error.

12.2.3 The types of errors listed in 12.2.1 and 12.2.2 above are generally human errors in programming instruments or controller generated commands which produce out of bounds conditions. The recovery from these errors normally involves human intervention and is beyond the scope of these recommended practices.

12.3 Error Correction.

Error recovery is often a required capability in systems having error detection. Specific error recovery techniques are beyond the scope of this recommended practice. Two common approaches are retransmission of data received in error and use of forward error correction (for example, Hamming codes).

Appendixes

(These Appendixes are not a part of IEEE Std 728-1982, IEEE Recommended Practice for Code and Format Conventions for Use with ANSI/IEEE Std 488-1978, IEEE Standard Digital Interface for Programmable Instrumentations.)

Appendix A
Message Format Examples: Syntax Diagram Uses

A1. General Comments

The conventions presented throughout this recommended practice provide the designer and user with a set of recommended formats for each different message type.

Rigid and mandatory message structures are considered to be too restrictive for broad general use. Typical formats for several different messages are given in this Appendix to illustrate the application of the recommended conventions. It is expected that individual product designs may deviate from the recommended conventions in special circumstances. Adherence to the recommended formats if encouraged strongly to increase the level of information interchange and compatibility among interconnected devices.

A1.1 Measurement Messages. The Figures in this section are in part similar to Figs 21(a) and 21(b).

A1.1.1 One voltage measurement of +10.002 VDC is taken on the 10 V range and expressed in NR3 (exponential) notation. The numeric value is preceded by an alpha header to indicate the type (DC) measured result. The message is concluded with CRLF for one measurement result.

Measurement Result:

DC+10002.E-03 CRLF

Syntax Construction:

Fig A1

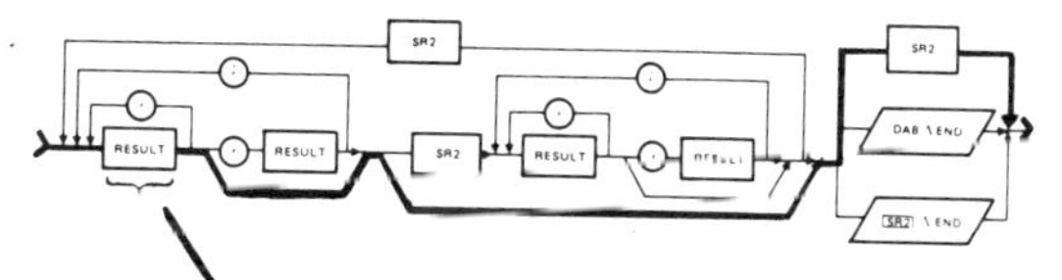

Pass 1
Separator CRLF

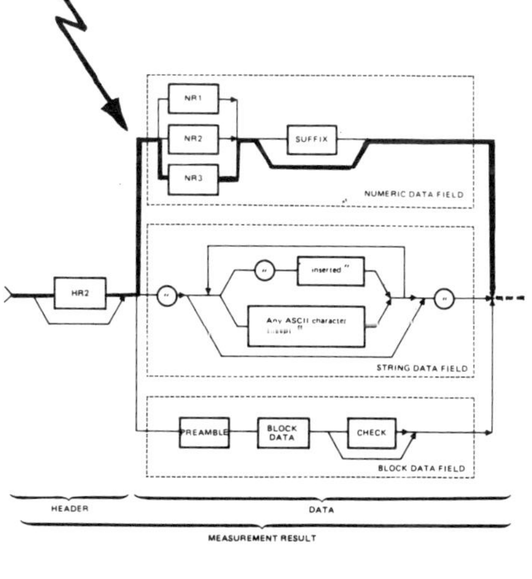

Pass 1
Header DC
Data+10002.E-03

Fig A2

A1.1.2 A spectrum analysis is made over a specified frequency span yielding 1000 measurements expressed in NR2 notation and separated by commas (in this case the dBm units are implicit).

Measurement Results:

−10.5,−11.1,(...)−86.1,(...)−11.1 [1 Λ END]

NOTE: The (...) symbols do not appear in the actual message. They serve only to indicate additional data in this example. [ ] indicates END sent concurrent with last byte [1] of *result*.

Syntax Construction:

Fig A1

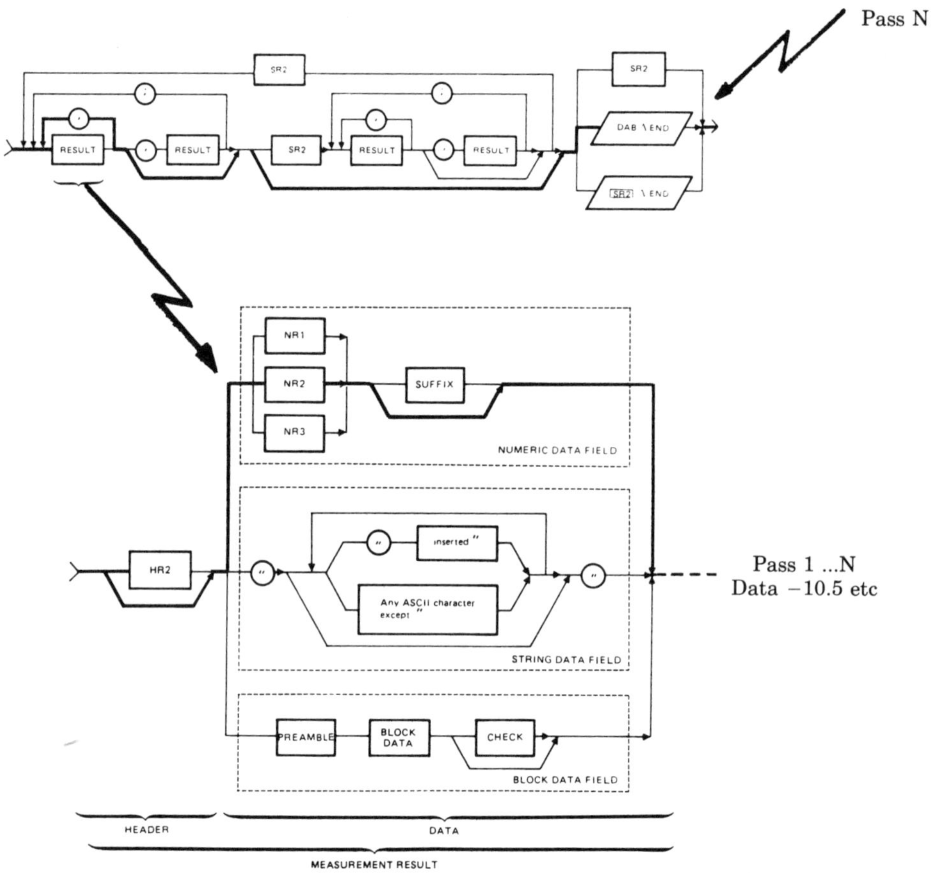

Fig A2

A1.1.3 A frequency counter with two channels (A and B) measures frequencies of 4.23 kHz and 2.60 kHz.

Measurement Result:

AFKHZ4.23,BFKHZ2.60 NL

Syntax Construction:

Fig A1

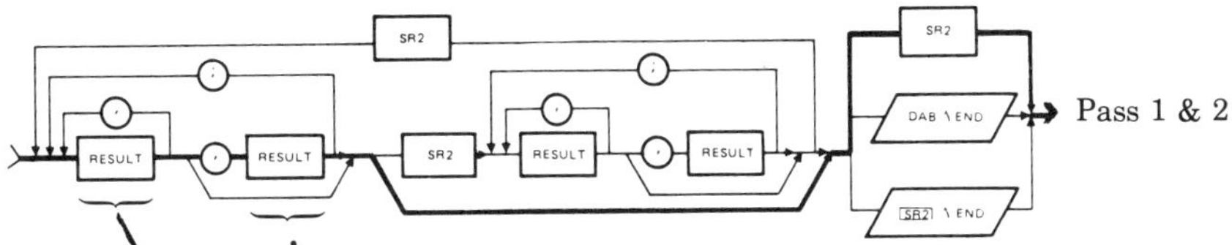

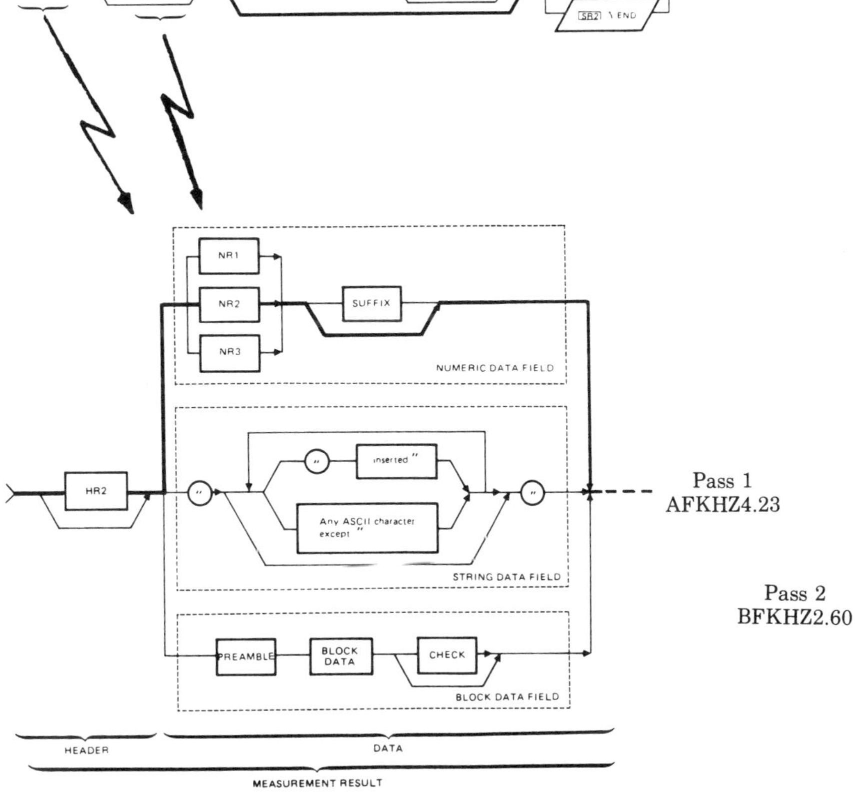

Fig A2

A1.1.4 A DVM takes a measurement of 10.002 V dc and sends it in NR2 notation.

Measurement result:

VOLTS 10.00[2 ΛEND]

Syntax Construction:

Fig A1

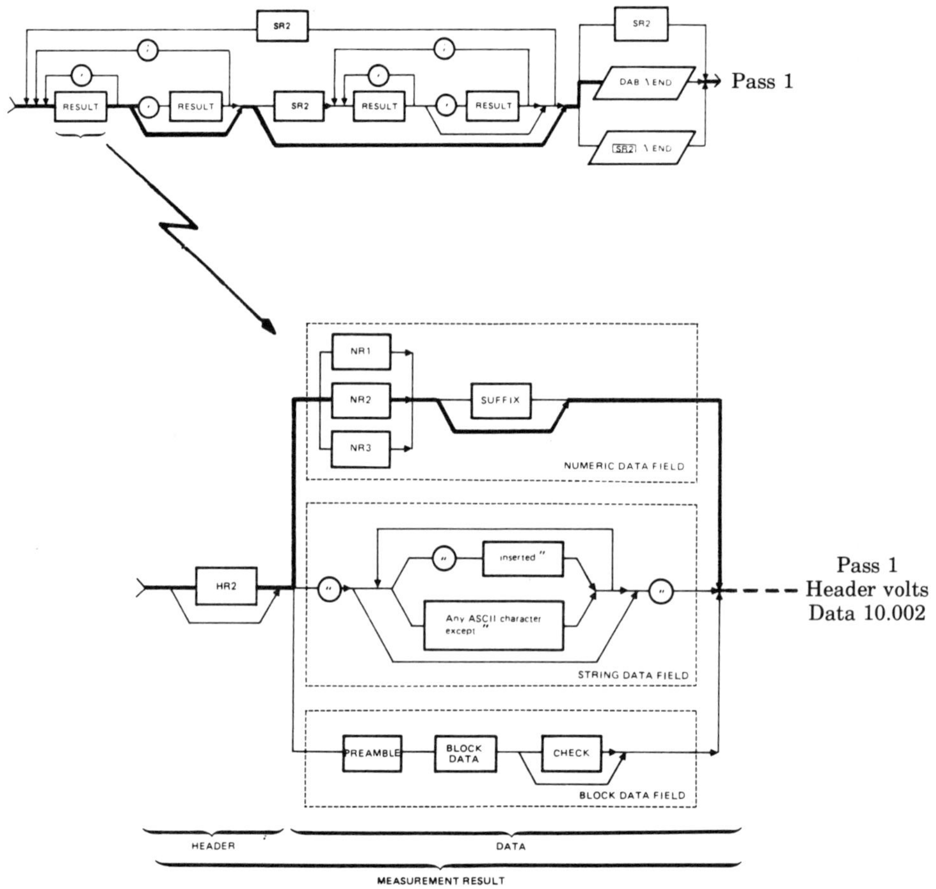

Fig A2

A1.1.5 An oscilloscope sends a waveform identification as a string value and the waveform itself (1024 data points) as a block value (requires two passes through the syntax diagram separated by | , |).

Measurement Results:

WAVEFORM "CHANNEL 1",
#CLLD . . . DDD(CRC1)(CRC2)

where:

L = 0000 0100 (length byte 1)
L =0000 0010 (length byte 2) (total length 1026 bytes)
D =0110 0000 (point 1)
: = :
D =0110 1000 (point 1022)
D =1011 0000 (point 1023)
D =1000 1101 (point 1024)
CRC1 =xxxx xxxx (CRC check byte 1)
CRC2 =xxxx xxxx (CRC check byte 2)

Syntax Construction:

Fig A1

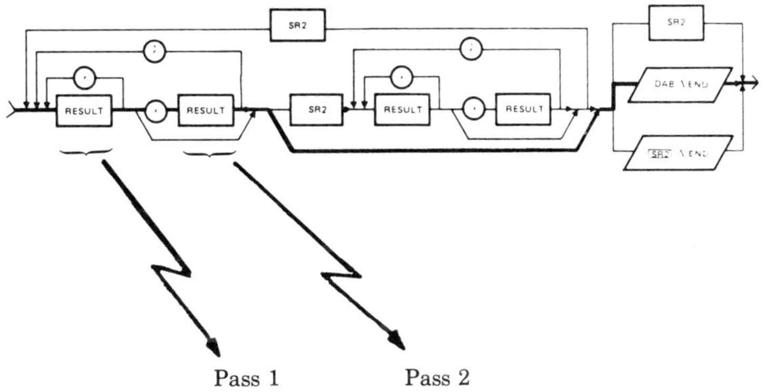

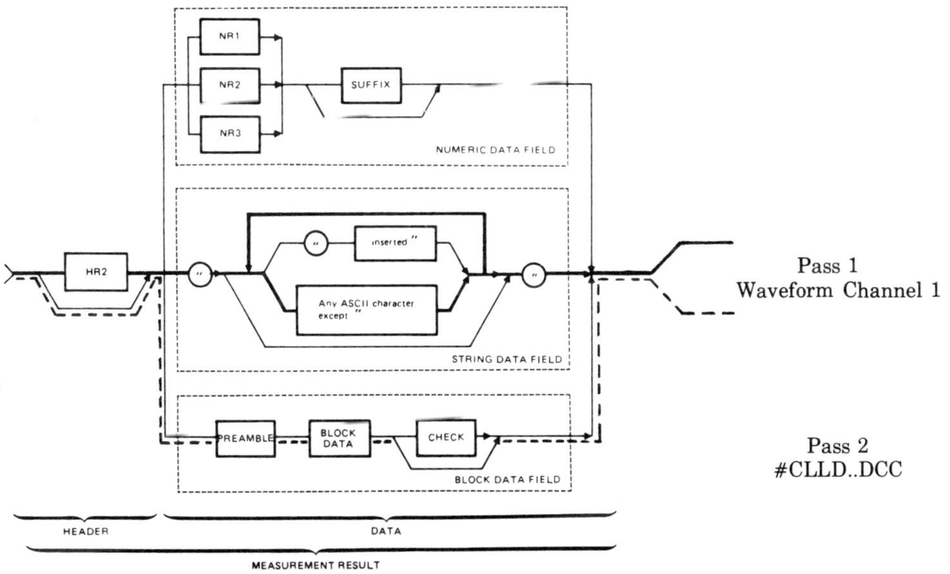

Fig A2

A1.2 Program Messages. The Figures in this section are in part similar to Figs 24, 25(b) and 25(c). (See Fig A3 and A4)

A1.2.1 A voltmeter is programmed to measure dc volts on the 10 V range (R4) upon receipt of internal trigger(T1)and output the measured quantity once the program(P) is executed. The dc function (FO) is performed using mode 3(M3).

Program Instruction(s):

Case A: - FOR4T1M3P

Case B: - FCTNORNGE4TRIG1MODE3 PROG

Syntax Construction:

Fig A3

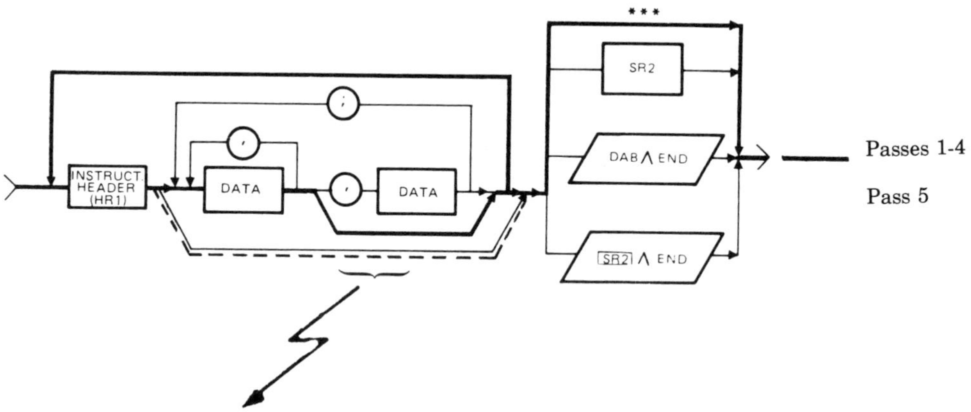

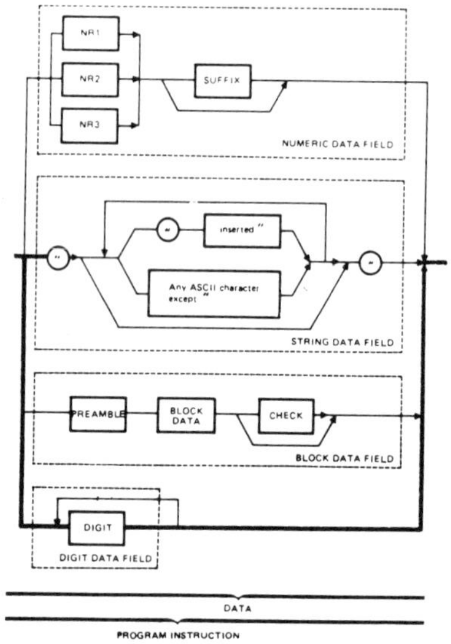

Fig A4

Pass	Case A Header	Case A Data	Case B Header	Case B Data
1	F	0	FCTN	0
2	R	4	RNGE	4
3	T	1	TRIG	1
4	M	3	MODE	3
5	P		PROG	

A1.2.2 A spectrum analyzer is programmed to make and store 1000 amplitude measurements using the A3 mode of storage and display. These measurements are to be centered about a frequency (CF) of 12.345 MHz over a span (SP) of 1000 Hz. The take sweep (TS) command causes this measurement sequence to be executed once.

Program Instruction:

CF12.345E+06HZSP1000HZTSA3

Syntax Construction:

Fig A3

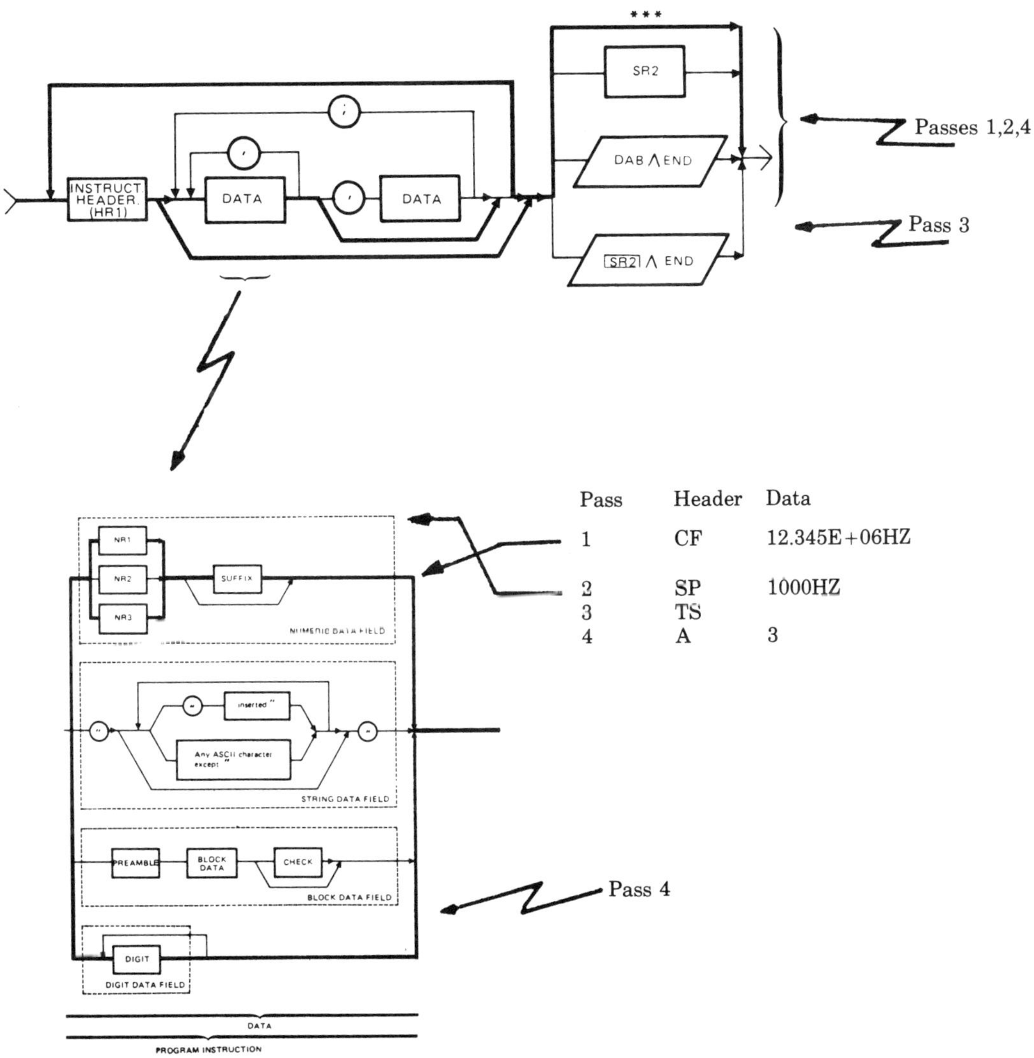

Pass	Header	Data
1	CF	12.345E+06HZ
2	SP	1000HZ
3	TS	
4	A	3

Fig A4

A1.2.3 A dual power supply is programmed to deliver both + and − 5.25 V at current limits of 120 MA and 60 MA respectively.

Program Instruction:

VI+5.25,120E−03;−5.25,60E−0[3 Λ END]

Syntax Construction:

Fig A3

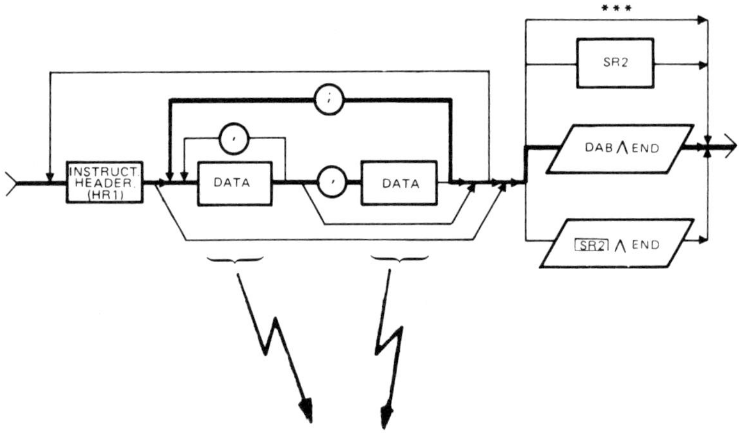

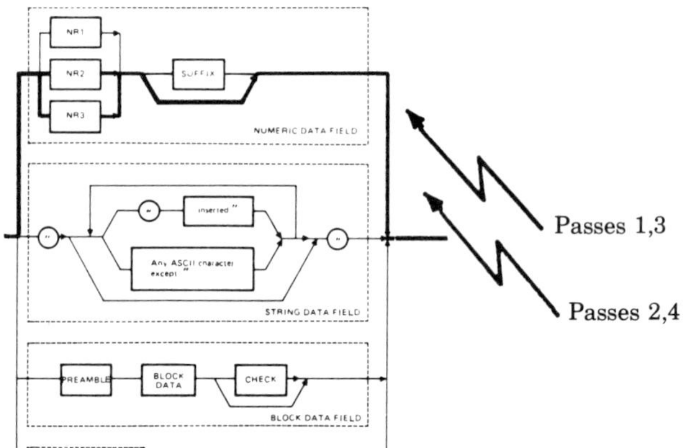

Pass	Message
1	+5.25
2	120E−03
3	−5.25
4	60E−03

(NOTE: SR3 (3 Λ END) is used as a separator)

Fig A4

A1.3 Program Messages. The Figures in this section are in part similar to Figs 24, 25(b) and 25(c). (See Figs A5 and A6)

A1.3.1 Set a DVM to 10 V range:

Program Instruction:

RANGE 10[V∧END]

Syntax Construction:

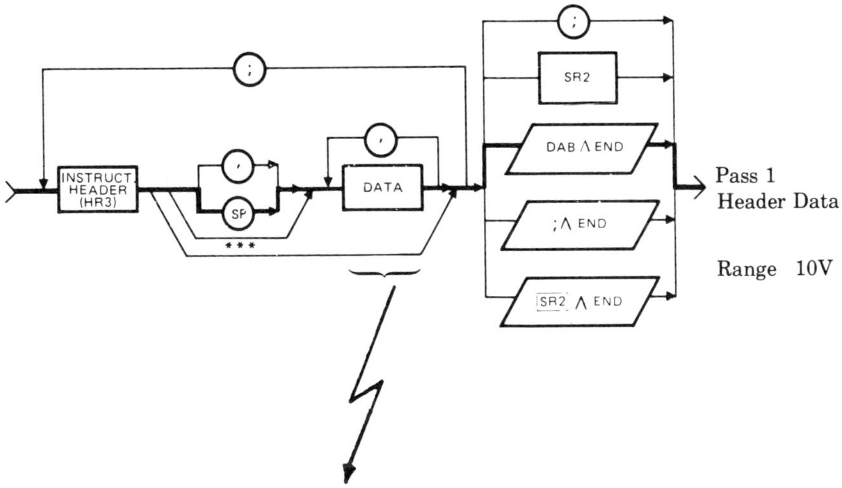

Fig A5

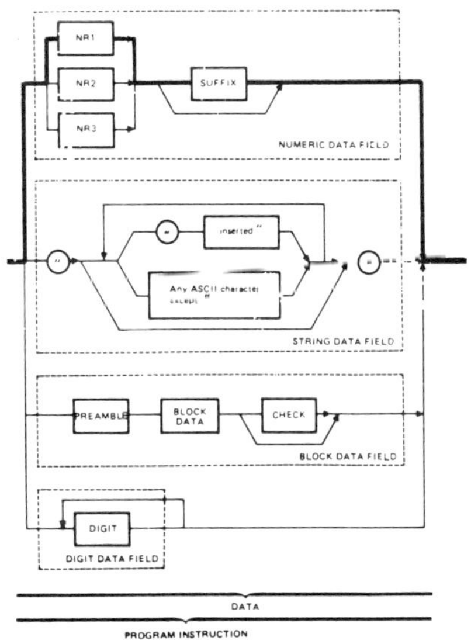

Fig A6

A1.3.2 Set a power supply to 10.5 V with the output on (requires two passes through syntax diagram).

Program Instruction:

VOLTS 10.5; OUTPUT ON <u>CRLF</u>

Syntax Construction:

Fig A5

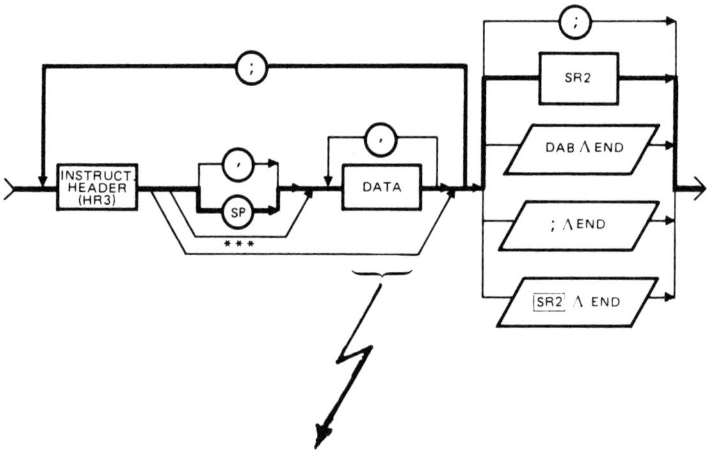

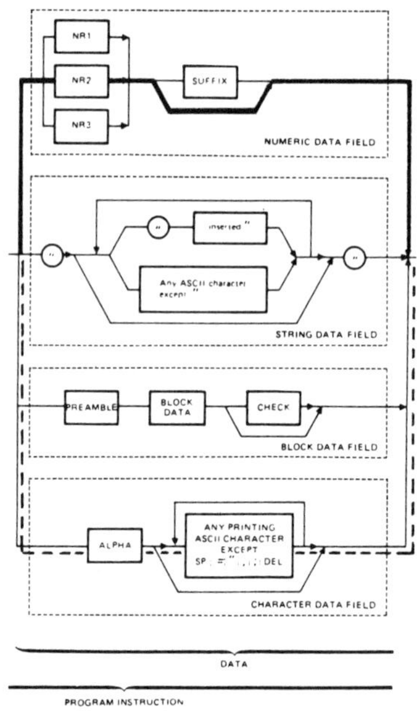

Pass	Header	Data
1	VOLTS	10.5
2	OUTPUT	ON

Pass 1
Pass 2

Fig A6

A1.3.3 Have a plotter draw a square and label it "DEMO".

Program Instruction:

MOVE 10,10 NL DRAW 90,10; 90,90; 10,90; 10,10 NL MOVE 40,0 NL
LABEL "DEMO["ΛEND]

Syntax Construction:

Fig A7

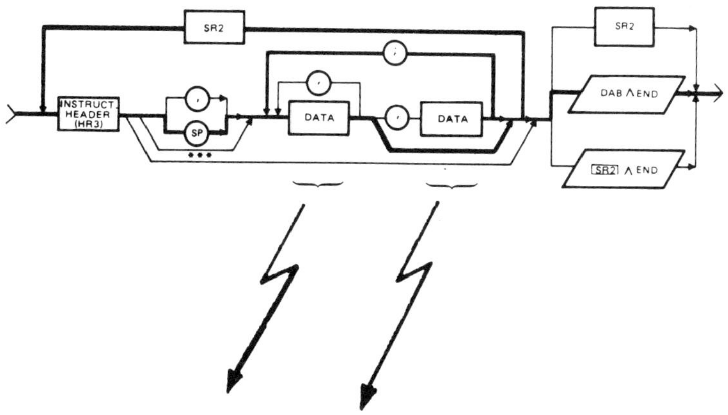

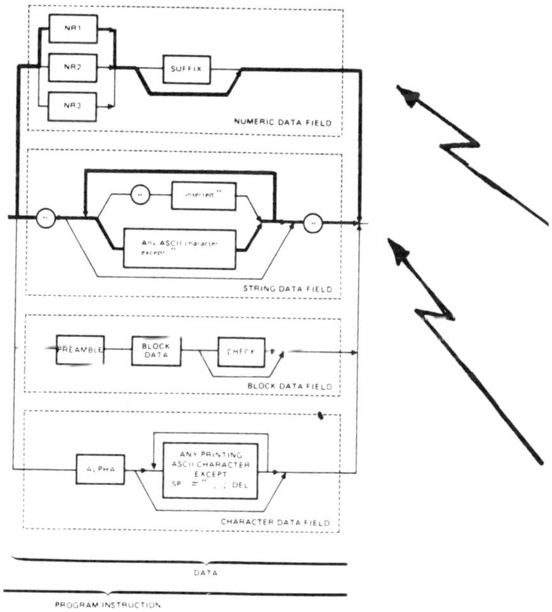

Pass	Header	Data
1	MOVE	10
2		10
3	DRAW	90
4		10
5		90
6		90
7		10
8		90
9		10
10		10
11	MOVE	40
12		0
13	LABEL	"DEMO"

Fig A8

A1.3.4 A plotter sends the coordinates of its pen position.

Program Instruction:

XPOS 53.68;YPOS 98.62:

Syntax Construction:

Fig A5

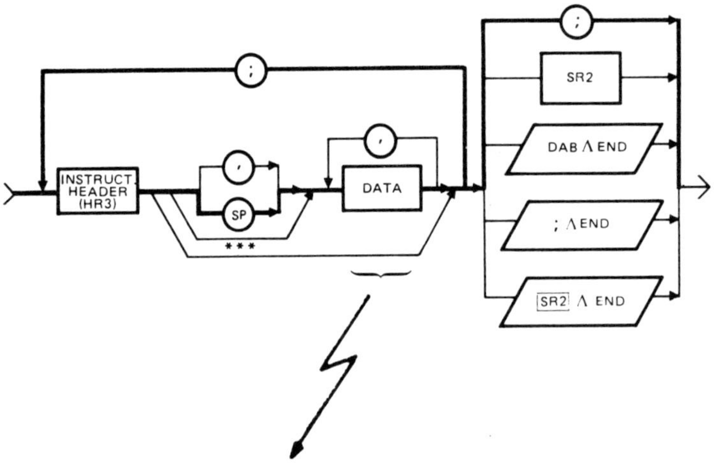

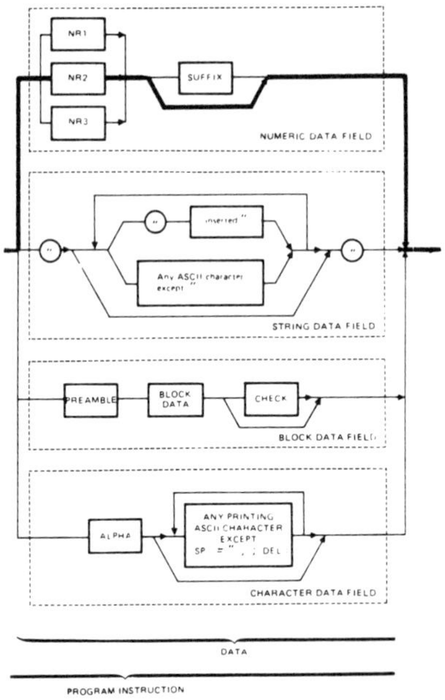

Pass	Header	Data
1	XPOS	53.68
2	YPOS	98.62

Passes 1,2

Fig A6

Appendix B

Format Capability Identification Codes

B1. General Comments

Sections 5 and 6 of this recommended practice define the syntactical structure of two broad types of messages for general use in the design and application of instrumentation systems; measurement messages and program messages. It is helpful to have products using this recommended practice employ the format capability codes (mnemonics) listed in Section B2 as a means to assist the user in identifying the specific syntactic structures employed within a particular product. The mnemonic representation of specific formats used by a product may be placed on that product's data sheet, operating manual and, where feasible, on the product itself.

B2 Format Capability Identification Codes

Table B1 lists those data fields and message unit constructs (particularly the hierarchial message constructs) and the assigned mnemonics considered to be useful in identifying product message format capability. Sections 5 and 6 of this recommended practice define the hierarchial message structures used when multiple message units are assembled together. Mnemonics for these constructs are also given.

Message Unit Formats and Hierarchial Separators:
Measurement Messages
o syntax per Figs 21(a) and 21(b) MM1
Program Messages
o syntax per Figs 23 and 26 PM1
o syntax per Figs 24 and 25(b) PM2
o syntax per Figs 24 and 25(c) PM3

Table B1
Data Field Mnemonics

Data Field	Elements		Mnemonic
Numeric data field			
Implicit point			NDR1
Fixed point			NDR2
Scaled (floating point)			NDR3
String data field			STDF
Character data field			CHDF
Block data field			
Binary Block,	No check,	Length specified	BDFA
Text Block,	No Check,	Length specified	BDFT
Binary Block,	No Check,	Length specified	BDFB
Text Block,	Checksum,	Length specified	BDFU
Binary Block,	CRC,	Length specified	BDFC
Text Block,	CRC,	Length specified	BDFV
Binary or Text,	No Check,	Unspecified Length	BDFI
Binary or Text,	Checksum,	Unspecified Length	BDFJ
Binary or Text,	CRC,	Unspecified Length	BDFK
Single Precision			BDFS
Double Precision			BDFD
Hexadecimal			BDFH
Octal			BDFO
Binary			BDFY
Digit data field			DIDF

In addition, it is considered useful to identify the set of explicit separators used with the product. Identification of the intramessage separators and EXIT separators is useful for both the messages sent and received, as appropriate.

B3 Application of Format Capability Codes

The mnemonic codes presented in this Appendix allow concise expression of the message syntax used by that device. *It is not expected that all possible combinations of a device's message syntax capability can be identified in short mnemonic form.* It is however, recommended that the operating manual describe the message syntax capabilities in detail such that assembly and use of the product is facilitated.

B3.1 Measurement Message Examples. Measurement message examples given in Appendix A are further illustrated here by the addition of the relevant format capability codes.

<table>
<tr><th>Message</th><th colspan="3">Format-Capability Codes</th></tr>
<tr><td>DC+10002.E−03CRLF</td><td>NDR3</td><td>MM1</td><td>CRLF</td></tr>
<tr><td>−10.5,−11.1,−86.[1ΛEND]</td><td>NDR2</td><td>MM1</td><td>| , | END</td></tr>
<tr><td>AFKHZ4.23,BFKHZ2.60NL</td><td>NDR2</td><td>MM1</td><td>| , | NL</td></tr>
<tr><td>VOLTS 10.00[2ΛEND]</td><td>NDR2</td><td>MM1</td><td>END</td></tr>
<tr><td>WAVEFORM "CHANNEL 1",
#CLLDDDD(CRC1)(CRC2)</td><td>BDFC
STDF</td><td>MM1</td><td>| , | END</td></tr>
</table>

B3.2 Program Message Examples. Program message examples given in Appendix A are further illustrated here by the addition of the relevant format capability codes.

<table>
<tr><th>Message</th><th colspan="3">Capability Code(s)</th></tr>
<tr><td>FOR4T1M3P</td><td>NDR1</td><td>PM1</td><td></td></tr>
<tr><td>FCTN0RNGE4TRIG1
MODE 3PROG</td><td>NDR1</td><td>PM1</td><td></td></tr>
<tr><td>CF12.345E+06HZS
P1000HZTSA3</td><td>NDR3
NDR1</td><td>PM1</td><td></td></tr>
<tr><td>VI+5.25,120E−03;
−5.25,60E0[3ΛEND]</td><td>NDR2
NDR3</td><td>PM1</td><td>| , | ; | END</td></tr>
<tr><td>RANGE 10[V Λ END]</td><td>NDR1</td><td>PM2</td><td>END</td></tr>
<tr><td>VOLTS 10.5;OUTPUT
ONCRLF</td><td>NDR2
CHDF</td><td>PM2</td><td>| ; | CRLF</td></tr>
<tr><td>MOVE 10,10NLDRAW 90,10;
90,90; 10,10NLMOVE
40,0NLLABEL
"DEMO["ΛEND]</td><td>NDR1
STDF</td><td>PM3</td><td>| , | ; |
END | NL |</td></tr>
</table>

Appendix C
Preferred SI Units and Multipliers

C1. General Comments

The preferred SI units and multipliers are representative of the information presented in ANSI/IEEE Std 268-1982 Metric Practice, IEC Publication 27-1 and 27-2, Letter Symbols to be Used in Electrical Technology and in ISO 1000-1973, SI units and Recommendation for Use of Their Multiples and ISO 2955-1974, Information Processing Representation of SI and other Units for use in System with Limited Character Sets (†according to ISO 2955 revision now under consideration).

NOTE: The SI units and multipliers listed below are to be considered as general recommendations and guidelines only. The assignment of absolute semantic meaning to specific characters is beyond the scope of this Recommended Practice. The designer should select characters in such a way as to avoid ambiguity of meaning.

C1.1 Base SI Units

	Common Usage	Upper Case
meter	m	M
kilogram	kg	KG
second	s	S
ampere	A	A
kelvin	K	K
mole	mol	MOL
candela	cd	CD

C1.2 Derived SI Units

hertz	Hz	HZ
newton	N	N
pascal†	Pa	PAL
joule	J	J
watt	W	W
volt	V	V
farad	F	F
ohm	Ω	OHM
siemens	S	SIE
weber	Wb	WB
tesla	T	T
henry	H	H
lumen	lm	LM
lux	lx	LX
bel *	B	B

*Common use indicates decibel is preferred (dB)
†According to ISO 2955 revision now under consideration.

C1.3 Other Units

	Common Usage	Upper Case
grade (angle)	g(s)*	GON
degree (angle)	°(s)	DEG
minute (angle)	′(s)	MNT
second (angle)	″(s)	SEC
liter	l	L
are	a	ARE
minute (time)	min	MIN
hour	h	HR
day	d	D
year	a	ANN
gram	g	G
ton	t	TNE
bar	bar	BAR
poise	P	P
stokes	St	ST
electronvolt	eV	EV
Degree Celsius	°C	CEL
atomic mass unit	u	U

*(s) indicates symbol is used in the right superscript position (like an exponent).

C1.4 Multipliers

Multi-pliers	Factor by which Unit is Multiplied	International Symbol (Common Use Symbol)	Representation for Simple Character Sets
eta**	10^{+18}	E	EX
peta**	10^{+15}	P	PE
tera	10^{+12}	T	T
giga	10^{+9}	G	G
mega	10^{+6}	M	MA****
kilo	10^{+3}	k	K
hecto	10^{+2}	h	H
deca	10^{+1}	da	DA
deci	10^{-1}	d	D
centi	10^{-2}	c	C
milli	10^{-3}	m	M
micro	10^{-6}	u	U
nano	10^{-9}	n	N
pico	10^{-12}	p	P
femto	10^{-15}	f	F
atto	10^{-18}	a	A

****Use of M in place of MA is the preferred practice if no textual confusion results.
**According to ISO 2955

Appendix D
Listen Mode Recommendations

D1. General Comments

Devices are expected to send device-dependent data (in TACS) according to the message formats defined throughout this Recommended Practice. Both instruments and controllers can however, improve the level of information interchange by listening "forgivingly" when in LACS. To listen forgivingly, a device should accept not only messages that are in the exact formats specified in Sections 5 and 6, but also messages that may have minor variations of these format structures. Some of the likely variations are due to inadvertent human error and limited capability in some products to provide a variety of output formats resultant from cost or performance considerations (for example, an instrument optimized for one single application but used in a totally different application).

D2. Specific Recommendations

In order to make the use of instruments and controllers more flexible in a variety of system applications and to promote the highest level of information interchange possible with ANSI/IEEE Std 488-1978 products, these recommendations are given with the idea of listening "forgivingly".

(1) Numbers (that is NR1-NR3) having a ZERO value and negative sign should be accepted.

(2) NR3 representations that do not have a signed exponent should be accepted.

(3) If a device receives an NR that has greater precision than the device can handle internally, the device should round-off the value rather than truncate it.

(4) Wherever feasible, HR and other data sent to a device as lower case alpha characters should be interpreted as the corresponding upper case characters. This capability makes a device, instrument or controllers, easier to use where the sending device outputs either upper or lower case alphas as part of their ordinary operations.

5. It is recommended that a listener function not hold its Acceptor Function in ACDS by asserting and holding NDAC. A listener needing to suspend bus transfers should do so in ANRS by holding NRFD. Additional guidelines may be found in ANSI/IEEE Std 488-1978, 5.3.2. It is to be noted that the SRQ message is an asynchronous event and is not to be construed as an interrupt when a device function fails or needs attention by the controller. A device that expects a fast response to its SRQ message from the controller must so indicate on the relevant documentation. Use of SRQ in this manner complicates the configuration and ease of use of the system and is not recommended.

Appendix E

E1. Floating Point Representations

The block representations specified throughout 4.3.2.3, 5.2.1.3.3 and 6.2.1.3.4 reference means to express both single and double precision floating point numbers.

An IEEE draft standard is under consideration which addresses this subject and contains specific proposals for single and double precision representations as well as extended single precision representation considered useful for application to ANSI/IEEE Std 488-1978 compatible products.

E2. Extracts from Draft Standard Floating-Point Arithmetic[3]

E2.1 Basic Floating Point Formats. Any nonzero real number may be expressed in normalized floating point form *as* $+/-2^{e*} f$, where e is the signed integer exponent and the significant digit field *f* satisfies $1 \leqslant f <$. The draft standard describes a machine representation of a finite subset of the real numbers based on this floating-point decomposition, and prescribes rules for arithmetic on them.

There are two basic formats, single and double precision that may be implemented in one of the combinations shown in Table E1. Single is required since it is useful as a debugging precision and is efficient over a wide range of applications where storage economy matters.

A normalized nonzero number X in the single format has the form

$$X = (-1)s^{*}\ 2^{E-12.7*}\ (1.F) where$$

S = sign bit
E = 8-bit exponent biased by 127
F = Xs 23-bit fraction which, together with an implicit leading 1, yields the significant digit field "1.–".

The values 0 and 255 of E are reserved to designate special operands discussed in later sections; one of them, signed zero, is represented by E=F=0. Normalized nonzero single numbers can range in magnitude between 2 −126* 1.000...00 and 2 −127* 1.111...11, inclusive.

The number X above is represented in storage by the bit string:

S	E	F

This encoding has the special property that the order of floating point numbers concides with the lexicographic order of their machine counterparts when interpreted as sign-magnitude binary integers, facilitating comparisons of numbers in the same format.

[3]Extracts are taken from P754, Binary Floating Point Arithmetic, now in preparation. It is not an approved standard.

Table E1
Basic Floating Point Formats

	Single	Double
Fields and widths in bits:		
S = Sign	1	1
E = Exponent	8	11
L = Leading bit	(1)	(1)
F = Fraction	23	52
Total Width	(1)+32	(1)+64
Sign:	+/−represent by 0/1 respectively	
Exponent:	biased integer	
Max E	255	2047
Min E	0	0
Bias of E	127	1023
Normalized numbers:		
Range of E	(Min E + 1) to (Max E −1)	
Represented number	$(-1)^{s*}\ 2^{E\ bias} * L.F)$	
Signed zeros:		
E	Min E	Min E
L	(0)	(0)
F	0	0
Reserved operands:		
Denormalized numbers:		
E	Min E	Min E
L	(0)	(0)
F	nonzero	nonzero
Represented number	$(-1)^{s*}\ 2^{E\ bias}\ (L.F)$	
Signed ∞ S:		
E	Max E	Max E
L	(0)	(0)
F	0	0
NaNs:		
E	Max E	Max E
L	(0)	(0)
F	nonzero	nonzero

F = system-dependent, possibly diagnostic, information.

E2.2 Extended Formats. To perform the arithmetic operations on numbers stored in the single and double formats, a system will generally unpack the bit strings into their component fields S,E, and F. Moreover, the leading significant bit will be made explicit, and perhaps the bias will be removed from the exponent. An explicit recommendation for either single or double extended precision is beyond the scope of the recommended practice.

E2.3 Relationships to DI01-DI08 Signal Lines. Transmission of floating-point representations, according to ANSI/IEEE Std 488-1978 shall be structured in accordance with the following relationships between DIO signal lines and the floating-point numbers. In this recommended practice MSB and LSB refer to most/least significant bit.

E2.3.1 Single Precision

```
8     1 8     1 8   1 8   1  <- DIO
----------------------------------
*S* E * *E* F * * F * * F *
----------------------------------
          ^                 ^
         MSB               LSB
```

E2.3.2 Double Precision

```
8   18   18  18  18  18  18  18  1 <-DIO
----------------------------------------
*S* E **E* F** F** F** F** F** F** F*
----------------------------------------
          ^                     ^
         MSB                   LSB
```

Index